EMERGENT INFORMATION

A Unified Theory of Information Framework

World Scientific Series in Information Studies — **Vol. 3**

EMERGENT INFORMATION

A Unified Theory of Information Framework

Wolfgang Hofkirchner

Vienna University of Technology, Austria

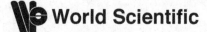 **World Scientific**

NEW JERSEY · LONDON · SINGAPORE · BEIJING · SHANGHAI · HONG KONG · TAIPEI · CHENNAI

Published by

World Scientific Publishing Co. Pte. Ltd.

5 Toh Tuck Link, Singapore 596224

USA office: 27 Warren Street, Suite 401-402, Hackensack, NJ 07601

UK office: 57 Shelton Street, Covent Garden, London WC2H 9HE

British Library Cataloguing-in-Publication Data
A catalogue record for this book is available from the British Library.

World Scientific Series in Information Studies — Vol. 3
EMERGENT INFORMATION
A Unified Theory of Information Framework

ISBN 978-981-4313-48-3

Printed in Singapore.

Preface

Fifteen years have passed since Peter Fleissner and I published an article in BioSystems titled "Emergent Information. Towards a unified information theory" [Fleissner et al. 1996]. At that time we had completed a research project on the genesis of information structures in which we examined what we considered the fundamental concept of informatics in co-operation with Klaus Fuchs-Kittowski. In the same year, we hosted the second conference on the Foundations of Information Science at the Vienna University of Technology. The motto we proposed was "The Quest for a Unifying Theory of Information" (see the proceedings [Hofkirchner 1999]). Our contribution was an appeal to the scientific community to combine efforts to overcome what we deemed an unsatisfying state of the art.

I then delved deeper into the issue and published a monograph in German in [2002]. This was prompted by the realisation that attempts to hypothesise or theorise about information in a unifying manner was not mainstream and that many scientists disbelieved in the feasibility of a single generic concept of information. We had already realised that a Unified Theory of Information (UTI) would need a unified theory of self-organisation too. I had the opportunity to contribute, together with John Collier, Rainer E. Zimmermann and others, to developing positions of an "Evolutionary Systems Theory" (EST) in another research project on "Human Strategies in Complexity" that ran from 2001 to 2004. Zimmermann instigated my ideas on the philosophical underpinning of UTI and EST, which I called Praxio-Onto-Epistemology (POE).

During my stay at the Paris-Lodron University of Salzburg from 2004 to 2010 I began to elaborate my positions in UTI, EST and POE. I owe thanks to my working group, all of who have become members of the UTI Research Group I had founded in 2003, and to numerous students. This book wrap-ups these elaborations. I found time to write this book thanks to a sabbatical granted by the University of Salzburg, a stay as Visiting Professor at the University of León in fall 2009 as well as a visit to the Open University of Catalonia in Barcelona in spring 2010. The book draws upon, but is a considerable extension and revision of, previous publications [2009, 2010a]. Intensive discussions with Josè María Díaz Nafría and Francisco Salto Alemany helped me clarify and rework several ideas. I am grateful to Michael Stachowitsch for polishing the English of the manuscript.

The present book focuses more on clarifying my own approach than on discussing similar or divergent ones. I only touched upon the literature with which I could easily relate. Besides the scholars already mentioned, colleagues who influenced my approach the most include Edgar Morin, Ervin László, Werner Ebeling, Klaus Kornwachs, Klaus Haefner, Tom Stonier, Alicia Juarrero, Edwina Taborsky, Søren Brier, Claus Emmeche, and Bob Logan, to name but a few.

I chose the title "Emergent Information" to highlight the continuity with my earlier efforts. "Emergent Information" points to the fact that the information concept is still emerging in society and in the various disciplines. It also alludes to the sciences-of-complexity information concept, which I present to link information to self-organisation such that emergence is a *sine qua non* for information to be generated.

The stage has changed since 1996. In 2010 the first-ever international conference "Towards a New Science of Information" took place in Beijing. My approach is a specific proposal about what the theoretical foundations of a new science of information could look like.

Wolfgang Hofkirchner, Vienna, March 2011

Contents

List of Tables

List of Figures

Part 1

Towards a Science of Information

Chapter 1

The Dawn of a Science of Information

There is a revolution coming. It will not be like revolutions of the past. It will originate with the individual and with culture, and it will change the political structure only as its final act. It will not require violence to succeed, and it cannot be successfully resisted by violence. It is now spreading with amazing rapidity, and already our laws, institutions and social structure are changing in consequence. It promises a higher reason, a more humane community, and a new and liberated individual. Its ultimate creation will be a new and enduring wholeness and beauty – a renewed relationship of man to himself, to other men, to society, to nature, and to the land.

The revolution is a movement to bring man's thinking, his society, and his life to terms with the revolution of technology and science that has already taken place. Technology demands of man a new mind – a higher, transcendent reason – if it is to be controlled and guided rather than to become an unthinking monster.

– Charles E. Reich, The Greening of America, 1970 –

Currently, a Science of Information does not exist. What we have is Information Science. Information Science is commonly known as a field that grew out of Library and Documentation Science with the help of Computer Science: it deals with problems in the context of the so-called storage and retrieval of information in social organisations using different media, and it might run under the label of Informatics as well. A Science of Information, however, would be a discipline dealing with information processes in natural, social and technological systems and thus have a broader scope. This is how the term Information Science is understood by a community of academics from different fields of science, engineering, humanities and arts who have been gathering around a conference series, a mailing list and a website with the

3

abbreviation "FIS" (Foundations of Information Science) for more than a decade.

Currently, no (Unified) Theory of Information is available. Information Theory exists: it is a branch of mathematics and engineering science inaugurated by Claude E. Shannon's paper "A Mathematical Theory of Communication" [1948]. That contribution dealt with the problem of keeping the signal–noise–ratio in communication transmission channels (hence the designation "channel" model) under control – a war-related problem. The entity that Shannon's formalism intended to measure soon came to signify "information". A (Unified) Theory of Information would have to deliberate about whether to generalise a framework that specifically focuses on syntactic aspects and omits semantic ones, yet reaches out to fields that require referring to semantics. This is precisely the brunt of many criticisms ever since. The result has been alternative theoretical approaches, none of which ever succeeded in being recognised by the overall scientific community as a more general Theory of Information that would, in fact, incorporate Shannonian Information Theory.

Equally, no unifying scientific information concept is available. We are accustomed to living with a multiplicity of diverse, and even contradictory, concepts of information. These are used throughout the edifice of natural, social and human, and engineering sciences, not to mention everyday thinking. Attempts to unify information concepts run the risk of getting caught in reductionism or its "twin", projectivism. A convincing information concept would need to sidestep these traps.

A unifying information concept is the core of a Unified Theory of Information, which in turn is the core of a full-fledged Science of Information. Thus, the task of an as-yet-to-be-developed Science of Information is to study the feasibility of, and to advance, approaches toward a Unified Theory of Information and toward a unifying concept while constantly being aware of a potential failure of the project.

1.1 In the Tower of Babel

The fragmentation, heterogenisation, and disintegration of what today is understood by the term of information has historical preconditions.

1.1.1 *The rise and fall of "information"*

The usage and meaning of the notion "information" has changed over the course of history. Drawing upon Rafael Capurro's seminal work of [1978], which is the ultimate source of reference, as well as on his later works on Angeletics, e.g. [2000, 2003], it appears that there has been an upward trend in the usage of the notion since Antiquity, that is an increase in incidence in the West. This was accompanied by a downward trend as to the meaning of the notion, that is a narrowing down of the range of objects signified and thus a thinning out of the content[a].

The roots of the notion of information lie in the beginnings of our civilisation – in Greek-Roman Antiquity. The Latin noun "informatio" and the Latin verb "informare", because of its root syllable "forma", refer back to the ancient Greek notions "typos", "morphe" and "eidos/idea", which mean the "stamped" or "stamping", "shape", "gestalt", "form". The verb "informare" therefore means "to shape something", "to give form to something", "to bring something into form".

Antiquity and the following (scholastic) Middle Ages knew several connotations of the verb. They manifest a great variety of meanings depending on what the subject and what the object of the activity might be (Table 1.1).

---

[a] This conclusion I draw is not in accordance with that of Capurro. He does not adhere to certain assumptions made in his early work. I base my interpretation on his original findings.

Table 1.1. Meanings of "informare"

subject	object	activity
	man	to educate
man	nature	to design, to construct, to craft
	man	to imprint something on the soul (in the epistemological sense)
nature	nature	to produce itself
	man	to teach, to instruct
god	nature	to shape

The following meanings can be distinguished:

(1) if the subject was man
   (a) and the object was man, then "informare" signified the activity of educating (in the broad sense of the antique pedagogical meaning that included the shaping of morality through role models);
   (b) and the object was nature, then the term meant designing, constructing, crafting (as in handicraft);

(2) if the subject was nature
   (a) and man its object, it meant the activity of imprinting in the sense of gaining knowledge (natural things are mapped to the soul according to their form);
   (b) and the object was nature itself, then it referred to the activity of producing itself (e.g., an organism is produced by nature);

(3) if, finally, god(s) took the role of the subject, then the activity was
   (a) teaching and instructing in the case of man as object
   (b) and shaping in the case of nature as object (giving form to substance).

Accordingly, the corresponding noun "informatio" had two different fundamental meanings:
(1) that of the activity of giving form to something/someone, and
(2) that of the result of this activity – being given form by something/ someone.

Note that meaning (2b) is an astonishingly modern-sounding notion. If we view humans as part of nature and assume that no god intervenes in natural processes, then this notion anticipates the very concept of self-organisation of today.

The whole variety of different meanings presented so far have one thing in common: in-forming means a process whose result is the emergence of something new.

At the dawn of the Industrial Age, the information concept spread from the Latin language into national languages and assumed an everyday content that did not cover all medieval meanings. With respect to pedagogy, the humanistic aspect of information as the education and development of human personality for the good and beautiful was set aside by the advent of modern philosophy. The rationalistic aspect of information was stressed as the intellectual process of communicating knowledge. In law, the information concept was at that time used to signify investigations.

The above developments cleared the way for the ultimate narrowing of the concept's meaning from signifying the process of communication or investigation to the exclusive signification of what is being communicated or investigated. This truncation of the information concept was executed with the help of Shannon's findings and writings. It reduced "information", finally, to signify everything that is transmittable via communication technology. This led to the situation that "information" began to be used to designate something that, originally, in ancient Greece, was named "aggelia", which is "message". This had had a celestial origin and had been communicated by messengers and "angels" down to the humans on earth. It has now undergone secularisation, referring to the horizontal exchange between humans, moreover between humans and machines and even among machines.

The restriction of the information concept (along with its reification, i.e. the belief that it represents a kind of thing: the easiest way to imagine that which is being communicated is that it represents a thing) was key to its triumphal diffusion into literally all fields and disciplines in the decades after Shannon.

That development, however, has been losing momentum.

1.1.2 *At the chaos point*

Today, the use of the term of information seems to have reached its climax. At the same time, with each step in the narrowing interpretation of the term, its meaning seems to become more dubious. The end of this trend of usage and meanings of the information concept might mark a turning point. This condensation could not have progressed much further. At this point of time we are witnessing an apparent increase of attempts to complement or overcome the channel concept.

1.1.2.1 *Disciplinary attempts*

Numerous criticisms have been levelled at the Shannon type of "information". That type has grown out of deliberations of single disciplines and points to particular problems. These problems mainly involve issues of how meaning enters the stage of information (a semantic question) and of how systems respond to information input (a pragmatic question).

The semantic information concept

The shortcomings of the new, narrow information concept were soon detected, with Donald M. MacKay being among the first to do so. Along with Shannon, he was invited to meetings in the legendary Macy Conferences series, which were famed for giving rise to cybernetics, system theory, and information theory. The eighth conference in March 1951 in New York City had "information as semantic" on the agenda. N. Katherine Hayles [1999, p. 74] reports that MacKay did not see

> too close a connection between the notion of information as we use it in
> communications engineering and what [we] are doing here... the problem here is

not so much finding the best encoding of symbols...but, rather, the determination of the semantic question of what to send and to whom to send it.

MacKay critiqued Shannon's equation of information and the signals transmitted and proposed, instead, to define information as

the change in a receiver's mindset, and thus with meaning

[Hayles 1999, p. 74]. Notwithstanding, Shannon's notion prevailed because it could be mathematicised while, at first sight, MacKay's could not. Shannon's elaboration of the information concept achieved the status of "Information Theory". Interestingly, as Robert K. Logan [forthcoming] annotates, it was MacKay who is cited in the Oxford English Dictionary as being the first to use the term in a heading in an article he published in the March 1950 issue of the Philosophical Magazine., Shannon's theory, in contrast, could best be described as a mathematical theory of signal transmission.

MacKay's insistence on including the meaning was not in vain. Another irony of history is that Gregory Bateson, one of the core group members of the Macy Conferences, anthropologist and cybernetician, has gained fame for his definition of information as "a difference which makes a difference" [1972, p. 453], while in the scientific community MacKay is credited with having defined information some years earlier in almost the same manner as "a distinction that makes a difference" (albeit without quote)[b].

Another early attempt at including semantics was undertaken by Yehoshua Bar-Hillel and Rudolf Carnap [1953], who unsuccessfully tried to develop a logical calculus. This line of thought was followed by Hintikka (see [Hintikka and Suppes 1970]) and Dretske [1981] and is still being pursued, e.g. by Luciano Floridi [2003, 2004, 2005].

The pragmatic information concept

Warren Weaver [1949], when writing an introduction to Shannon's ideas and admitting that Shannonian information was not about semantics, already gave the following classification:

---

[b] Floridi [2009, p. 32] states that "distinction", in comparison to "difference", has an epistemological rather than an ontological twist.

(1) there are technical problems concerning the quantification of information that are dealt with by Shannon's theory,
(2) there are semantic problems relating to meaning and truth,
(3) and there are what he thought to be "influential" problems that address the impact and effectiveness of information on human behaviour, which he thought had to play an equally important role.

This classification precisely describes what can be labelled the concept of "syntactic", "semantic" and "pragmatic" information, respectively.

The pragmatic information concept has been coined by Ernst Ulrich von Weizsäcker [1986] and his wife Christine von Weizsäcker "as if the receiver mattered": it is the receiver on which information has an impact; and it is the receiver who creates information. Information is said to be that which produces information. Different concepts have been developed, such as the trade-off between surprise and redundancy, intelligibility, context dependence, wholeness [Gernert 1996]. Klaus Kornwachs is one of the advocates and representatives of the pragmatic information concept today. This view is relevant for decision-making processes. Nonetheless, this view, which resembles behaviouristic black-box models when relating the input of a system to its output, raises questions about whether the reified concept has really been broken up.

1.1.2.2 *Transdisciplinary attempts*

Besides the attempts to complement the Shannon information concept from a particular point of view, there has been a search for a concept that can overcome the shortcomings by integrating the various aspects of information processes. The useful aspects, if any, of the Shannonian term should be included as a special case, when extending the restricted information theory into a new, universal theory.

Clearly, transdisciplinary undertakings that strive for the bigger picture tend to be affiliated to philosophy, like – from different angles – the works of Rafael Capurro, Luciano Floridi or Kun Wu, and cross-disciplines such as cybernetics, system theory, evolutionary theory. Unsurprisingly then, a precursor of this line of thinking was developed from a philosophical perspective on cybernetics and system theory – by the French philosopher and sociologist Edgar Morin. Already in 1977, Morin developed a concept of information that defines information as that which is functional for the maintenance of a system. Only the first volume of his six-volume oeuvre *La Méthode* was translated into English (*The Nature of Nature*) [1992]; the German translation followed in 2010.

Conceptualisations which date from the second half of the 1980s mark a new period. These are:

(1) the hypothesis of the control revolution, which James R. Beniger [1986] uses to draw parallels between the breakthrough to the information society and former revolutions in the course of life and culture;

(2) and the hypothesis of the evolution of information-processing systems put forward by Klaus Haefner in [1988] (and edited in [1992b], see also [1992a]). It views the information society as the ultimate result of the evolution of systems in the universe which are capable of generating and processing ever higher information.

These two outstanding contributions are the initial steps toward a single and comprehensive science of information.

Writings of scholars who have a cross-disciplinary background build upon the same train of thought. One example is the three-volume work of the Dutch expert in International Relations, Johan K. De Vree [1990], who develops a system-theoretical approach that starts with thermodynamic considerations. By doing so, he avoids the fundamental shortcoming of cutting society free from the material-energetic world (a mistake made by Niklas Luhmann). Another example is the science-of-information trilogy written by Tom Stonier [1990, 1992, 1995], a trained biologist who ultimately became Professor Emeritus for science and society at the University of Bradford. Stonier offers an evolutionary perspective of societal development up to the information age. Both scholars had been active in the Foundations of Information Science community at its onset in the early 1990s.

Klaus Fuchs-Kittowski [1997], from the German Democratic Republic, made a unique attempt to combine philosophical, cybernetic, biological, sociological, technological, and computer science thinking and then focus on the origin of information.[c]

A recent milestone is Søren Brier's book on Cybersemiotics [2008]. It spans the full breadth of library and information science, cognitive science, consciousness studies, communication studies, semiotics, cybernetics, systemics and informatics, evolutionary theory, ethology, sociology, epistemology, ontology and philosophy of science. It challenges, as the title of chapter 1 already reveals, "the Information-Processing Paradigm as a Candidate for a Unified Science of Information". Though the book has a misleading subtitle which insinuates that "information is not enough", Brier [p. 381] asserts his commitment to a modern systems thinking that

> views Nature as containing multilevel, multidimensional hierarchies of interrelated clusters, which together form a heterogeneous general hierarchy of processual structures: a 'heterarchy.' Levels emerge through emergent processes when new holons appear through higher-level organization.

He concludes [p. 381]:

> Across levels, various forms of causation [...] are more or less explicit (manifest). This leads to more or less explicit manifestations of information (sic!) and semiotic meaning (sic!) at the various levels of the world of energy and matter.

and states [p. 382]:

> Meaning is generated through the entire heterarchy [...]

This means that those levels that Brier classifies as protosemiotic and information-only levels are also, ulimately, perfused with signs.

Robert K. Logan, a physicist, worked with Marshall MacLuhan on technological development and published books on the evolution of languages up to the Internet. He revisited MacLuhan in the light of modern evolutionary systems thinking, and worked recently with Stuart Kauffman and others on biological information. The result was a new definition of information. On that occasion, Logan completed a manuscript for a book titled *What is Information?* [forthcoming] in which he extends the definition to other realms of reality. This new

---

[c] Most of his writings are available in German only.

definition was found during the search for a nonreductionistic explanation of evolution of living systems. Kauffman et al. [2008, p. 36]

> identify information, which we here call "instructional information" or "biotic information," not with Shannon or Kolmogorov, but with constraints or boundary conditions, and the amount of information will be related to the diversity of constraints and the diversity of processes that they can partially cause to occur.

In addition, several approaches aim at theories of a global brain (e.g. the Principia Cybernetica Project group around Francis Heylighen, see for instance [1995, 1997], from a cybernetics point of view) or of a collective intelligence ([Lévy 1997] – in French 1994 –, from a philosophical point of view). Others draw parallels between superorganisms and mankind [Stock 1993] or between biotic and cultural developments in general (see e.g. the living systems theory of James Grier Miller from [1978] and the article [Miller and Miller 1992] or Peter Corning's Synergism Hypothesis from [1983]). Still others share an evolutionary perspective without referring to biology (e.g. [Malaska 1991], [Artigiani 1991]).

A number of sciences seek to contribute to a bigger picture of information. The first is the physical sciences ([Haken 1988], [Ebeling1989], [Lyre 1998], [Baeyer 2003]). Seife [2006], for example, even uses the term "Science of Information" in the subtitle of his book, The biological and cognitive sciences are represented by Collier [forthcoming], Moreno [1998], Moreno and Ruiz-Mirazo [2007], the philosophy of mind by Juarrero [1999]J, the semiotic approach by Taborsky [1999], and the technological perspective by Mingers [1995, 1996]) – to cite but a few.

And – to complete the multiverse perspective – there are contributions from outside the usual Western thinking. Tamito Yushida, Member of the Japan Academy, Professor Emeritus at the University of Tokyo and the Chuo University, participated in a Japanese project on clarifying information concepts. In 1993 he coined the term "informatic turn", by which he links philosophy of science and semiotics considerations [2005][d]. Using the theory of autopoietic systems going back to Umberto Maturana and Francisco Varela [1980], Toru Nishigaki [2007] at the University of Tokyo developed "Fundamental Informatics" as a common basis for all branches of information studies, which he sees as being divided between information science or information engineering and socio-informatics. China has brought forward *An Introduction to Theoretical Informatics* [Li et al., 2010]. This English edition of a student textbook provides a general theory of information that deals with the relationship of matter, energy and information, conceives an "information energy", formulates three laws of information (the law of the non-conservation of information, the law of the increase of information with time, and the law of the unlimited growth of information) and applies these categories to biotic and social evolution. One of the authors, Zong-Rong Li, trained in Systems Science, is Professor in Computer Science and Technology and Vice Director of the Social Information Science Institute at the Huazhong University of Science and Technology in Wuhan. Clearly, many other Chinese scholars will appear on the scene once they have published in English.

Since 2010 a "Center for Science of Information" has been sponsored by the National Science Foundation (USA) as a Science and Technology Center. It focuses on integrated research "around three application thrusts: life sciences, communication, and knowledge extraction from massive datasets" [CSoI 2010, 5]. Although the Center aims to arrive at new measures of information and thus follows a formal science approach, it shows that a Science of Information that goes beyond the Shannon concept may have begun to gain recognition by academia. Publications that have paved the way are "Formal Theories of Information – From Shannon to Semantic Information Theory and General Concept of Information" [Sommaruga 2009], "Asymmetry: The Foundation of Information" [Muller 2007], or "Information and its Role in Nature" [Roederer 2005].

---

[d] Paper prepared for the author's first presentation in English to an international audience, namely the First World Congress of the International Federation of Systems Research 2005 in Kobe.

A key recent publication is the book "Theory of Information" by Mark Burgin [2010, ix] "that is aimed at finding ways for a synthesized understanding the information phenomena and at building a general theory of information encapsulating other existing information theories in one unified theoretical system." That book suggests such a concept.

All these transdisciplinary initiatives make an even stronger case for substantiating the apparently unsubstantiated use of the term "information" throughout sciences than earlier point-by-point criticisms. Combined, they form constituents of a huge melting pot of scientific exchange and dispute. They have helped heat up the debate to a point where a critical value seems to have been reached. This is equivalent to a chaos point: either the debate will cool down and things will remain as they were, or a leap in quality will emerge from the chaos and bring about change and new order.

Hence, a Science of Information is an idea whose time has come!

1.2 After the Information Revolution is Before the Information Revolution

At first glance, it may seem to be an intrascientific issue whether or not there is an attempt in the field of information to grasp the big picture and develop a shared theory that would help understand the full range of different manifestations of information processes in society and the world at large. In everyday thinking, people strive to connect unconnected experiences and even to reconcile irreconcilable experiences in order to arrive at a coherent overall view. This is reflected in the psychologically well-described tendency of ordinary people to avoid cognitive dissonance. Likewise, science is heading for consilience – a term that attracted interest when Edward O. Wilson published his book of the same title [1998]. This consilience is a unity of knowledge that allows for ever better explanations and predictions. It is accomplished by constructing new theories that include the findings of the old theories as a type of approximation and at the same time are able to explain and predict phenomena that were not covered by the old theories. Thus, science tends toward increasingly overarching theories, toward ever more generalising theories, toward increasingly universal theories. Unified theories address universal aspects by unifying the multiplicity of so far incoherent theories bound to particular levels. Unified theories belong to the intrascientific progress toward the universal[e].

This development goes beyond pure scientific curiosity. The precondition is that science is not restricted to an ivory tower, but is a social undertaking that satisfies social demands. Accordingly there is an extrascientific function all science has to fulfil – the betterment of social life and solving problems that arise from social practice. It should come as no surprise, then, that – on the threshold of the information age – science is concerned with information and that there is a quest for a Unified Theory of Information (UTI) (see [Hofkirchner 1999]).

The information age is normally defined as the age of information societies into which industrialised societies are transforming. This is evident in the spread of new information and communication technologies (ICTs). The industrial age, in turn, is defined as the age of industrial societies into which agricultural societies have been transforming worldwide. Each transformation is known as a revolution and all revolutions together are said to form the evolution of civilisation (Figure 1.1).

On the one hand, there is a lag of social-scientific development behind societal and technological development. Development in technology is not accompanied by an equally rapid growth in scientific insight, let alone foresight, as to the impacts of technology on levels of society other than that of technological organisation. Attempts to observe and understand the basic nature of this change are still secondary. The public use of the notion of "information society" has been reduced to denoting a society in which applications of modern ICT are widely spread in order to facilitate the handling of what commonly is called "information". A scientific understanding of this transformation has not had time to develop. There is currently no proper Science of Information or proper science of the information society.

---

[e] As Aristotle put it: science is about the universal (See De Anima 5, 417b23; An. Post I 18, 81b7-9; An Post I 8, 75b21-23; Met I 3 983[a]25-26).

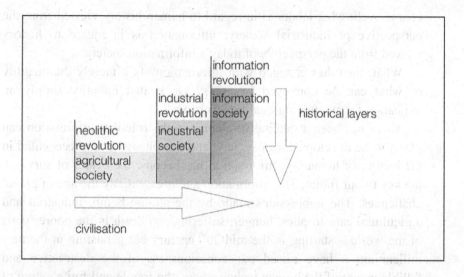

Figure 1.1. The standard view of human history as a sequence of scientific-technological revolutions transforming the whole of society.

On the other hand, the state of the relationship between science and modern technosocial development regarding information can, for example, be compared to the state Karl Marx was confronted with in respect to labour. In his time, labour could become and necessarily became a matter of scientific interest. This is because labour had gained a new role in society. It became something general in social life, that is, it was treated in society irrespective of its concrete characteristics, and thus became more abstract. Marx termed that a "real-abstraction" – an abstraction that occurred in reality due to the real treatment of labour in emerging capitalism. This became the basis for the general concept of labour in scientific thought. It was only then that the concept of labour could be stretched back to former social life in the history of humanity. At the same time, phenomena beyond industrial work could be subsumed under a common concept, the concept of labour, albeit as different manifestations. This notion of real-abstraction might lead us to assume that information has gained a sufficiently decisive role in today's society to foster a new scientific conceiving and theorising, i.e. that it has become a real-abstraction which is the rationale for devising a general idea as well: what labour is in regard to human history viewed from the perspective of industrial society, information is in regard to history viewed from the perspective of today's information society.

What, then, has changed with information? Is it merely the quantity of what can be conveyed by ICTs? Or is that quantity simply an indication for an ongoing qualitative change?

There has been a qualitative change in the role that information can play for the development of society, and this change is unprecedented in the history of humanity. Information has become the bearer of survival, the key to our future. The information age is essentially the age of global challenges. The impressions made by the atomic bomb, industrial and agricultural catastrophes, hunger, suffering and death in the poorer parts of the world – starting in the mid-20th century but persisting in the new millennium – have raised our consciousness of the destructive and fallible nature of the human technosphere, the fragile and finite nature of the human ecosphere, and the unsettled, unbalanced nature of the human sociosphere. It has become general knowledge that the existence of such global challenges can endanger the persistence of today's societies worldwide. The global problems are global in a twofold sense:

(1) they concern humankind as a whole (as object);
(2) they can also only be solved by humankind as a whole (as subject).

The risk this crisis carries is that humankind may be wiped out. The chance it offers, however, is that humankind may be raised to another level of humanity.

In 1995, the present author advanced the idea that these disparities in the development of the relations amongst humans, between humans and nature, and between humans and technology can be viewed as expressions of a human deficiency, i.e. an incapability to control and regulate the systems in question by information. The problems can be viewed as problems in controlling and regulating society, the environment and technology such as to ensure the maintenance of the systems and their functions critical for the survival of humanity [Hofkirchner 1995]. Five years later, this author suggested interpreting the problems as frictions in the functioning of the information processes within the above systems [Hofkirchner 2000].

Interestingly, the information revolution turns out to provide new potentials for reducing frictions. Floridi expounds that the "infosphere" will ultimately, by connecting systems to it one by one, turn into a "frictionless" cyberspace. He states [2007, 61],

> At the end of this shift, the infosphere will have moved from being a way to refer to the space of information to being synonymous with Being.

At the level of technology, the so-called ephemeralisation trend (ever increasing productivity and efficiency) plays a role: it is already decreasing friction and smoothening the processes in which matter, energy and information are said to circulate. According to Francis Heylighen, this holds true not only for natural and artificial systems, but also for social ones [Heylighen 2007, 9-10]:

> *Initially, interactions tend to be primordially competitive, in that a resource consumed □by one agent is no longer available for another one. In that respect, interactions are □characterized by social friction (Gershenson 2007), since the actions of one agent □towards its goals tend to hinder other agents in reaching their goals, thus reducing the □productivity of all agents' actions. Note that the two common meanings of the word □ "friction"—(physical) resistance, and (social) conflict—describe the same process of □unintended obstruction of one process or system by another, resulting in the waste of □resources.*

He continues [2007, 10]:

> *Like physical friction, social friction creates a selective pressure for reducing it, by □shifting the agents' rules of action towards interactions that minimally obstruct other □agents. Interactions, however, do not only produce friction, resulting in a loss of □resources, they can also produce synergy, resulting in a gain of resources.*

Thus, information proves the only remedy, given the malfunctions in society, the environment and technology that continue to aggravate the global challenges and to create obstacles to keeping society as a whole on a stable, steady path of development. Information is what is required to alleviate and reduce the frictions in the functioning of those systems that make up humanity – from the individual to ethnicities to nations to world society, from economy to polity to culture, from society to ecology to technology, from the social realm to the biotic realm to the physical realm. Information is what is required to steer society, to reorganise humanity onto a higher level of organisation. Thus, the continued existence of humanity may well be impossible without conscious and cautious interventions in its own development processes. This includes all spheres of intervention. This intervention is informational in nature, oriented toward relinking a world falling apart due to processes of heterogenisation, fragmentation and disintegration. It extends from human beings to all living organisms to matter, necessitating a deep understanding of all the information processes in the world we inhabit.

Today, knowledge as a capacity to act means the capacity to act *vis-à-vis* the global challenges. It means knowledge about how information guides the processes that put us at risk. Hence, information is the *conditio sine qua non* for the survival and development of humanity.

The "information revolution" is commonly held to be a revolution in science and technology. This notion dates back to the computer revolution that originated in World War II and has revolutionised all society thereafter. Today we are witnessing not only the convergence of computer technologies with telecommunication technologies and media technologies, but also the convergence with nanotechnologies from material science and with biotechnologies, amongst them genetics engineering, and with cognitive science. These are subsumed under the term NBIC. The "information revolution" is not over, is not past, but has turned out to be a permanent revolution, with the most dramatic changes still *ante portas*.

Awareness is growing that technological determinism is too myopic: the belief in a technological progress which *per se* entails social progress has diminished. Data, however, are not the *ultima ratio* of the changes to come, nor even is knowledge, regardless of its quantity. This contravenes the recent global hype of a "knowledge-based economy" and "knowledge society". Rather, it is wisdom which may make the emerging society a "wise society" – as a report of the High Level Expert Group of the European Union on *A European Information Society for Us All* stated more than a decade ago [HLEG 1997]. Such a society is capable of coping with the challenges arising from its own development. A scientific understanding of this transformation urgently requires a Science of Information.

1.3 Adding to the Critical Mass

The chaos point we have reached provides a window of opportunity for a paradigm shift. What scientific requirements need to be met when trying to develop a Science of Information? In other words, what is, in scientific terms, required by a future Science of Information [Doucette et al. 2007]? To be ready for the paradigm shift, we must anticipate now what will be required in the future.

Three contexts – each with its specific criteria – can be distinguished in terms of what makes science scientific:

(1) a context of application in which scientific knowledge is used to solve problems and is transformed into technologies, whether material or ideational;

(2) a context of justification in which scientific knowledge is critically exposed to possible refutations and corroborated in as far as it is not refuted, and theories are comparatively assessed;

(3) a context of discovery in which scientific knowledge is conjectured and theoretical assumptions are formulated in relation to empirical findings.

The first context concerns the objective, the end, the aims for which the scientific endeavour is undertaken. It fulfils the task of solving a problem that arises from practice in society and it is practically guided by the aims.

The second context refers to the object of the inquiry; this is the starting point that determines the domain, the scope to be researched.

The third context is about objectivation, i.e. methods that help objectivate or, substantiate, the findings; these strategies are the means to transport the scientific enterprise from the starting point to its end. This context is the tools used.

Accordingly, each science can be described in terms of (1) aims, (2) scope, and (3) tools.

A key assumption is that some relationship exists that connects aims, scope and tools: they build a three-levelled architecture with aims on top, scope at an intermediate position and tools at the bottom. Thus, despite a relative autonomy of each of the levels, aims define the possible scope (particular interests demand a particular section of reality to be selected for investigation), and the scope defines possible tools (not every method is appropriate to study a particular object). In reverse, specific tools determine the variety of the scope that is realised (the constitution of the object), and a particular scope determines the variety of possible aims to be served (the findings cannot be used for any practical guidance).

At each level, criteria apply that make scientific thinking distinct from (yet merely a thoroughly reflected continuation of) common sense:

(1) In the aftermath of the Positivism debate in German sociology between the Frankfurt School type Critical Theory and positivist Critical Rationalism, philosophy of science considerations seem to have agreed that value-free science does not exist. Clearly, technology serves humane values. Technology assessment must take into account the interests of those affected by technology and examine whether or not their needs, as well as general interests, are met. The dispute is mostly about which values to prioritise. Such a dispute requires explicitly formulating the values implicit in scientific and technological developments.

(2) Findings shall be anchored in theories which, preferably as universal implications, are statements describing general and necessary properties or general and necessary relations that cover the object of inquiry as a whole. As a rule, if a theory is challenged by a counterexample not obeying the law, another theory is sought to explain this phenomenon as well as all phenomena the old theory could explain. In times of normal science – to use the terminology Thomas S. Kuhn [1962] coined – no such need arises. Other times call for a scientific revolution (whose reach may differ according to the issues at hand) and for paradigms that compete with each other (which may range from different schools of thought within a subdiscipline to entirely contrasting world views).

(3) Logically, no induction compellingly leads from concrete empirical grounds to theoretical generalisations. Nonetheless, many methods have been devised by different disciplines to support the creative construction of theories. The scientific debate revolves around immanent and transcendent criticisms of whether or not the methods applied are appropriate.

What are the aims of a Science of Information?

What is the scope of a Science of Information?

What tools should a Science of Information make use of?

The proposal here is to determine the three levels of a desired Science of Information *ex negativo.* that is, in comparison or contradistinction to the features characterising the state-of-the-art of normal science that conducts information research. This is because they do not meet the needs of societal development today (Table 1.2).

Table 1.2. Aims, scope and tools in normal science information studies and in a Science of Information paradigm.

	normal science information studies		a Science of Information paradigm
aims	applied research: any feasible information concept	basic research: any imaginable information concept	use-inspired basic research: information concepts cast for meeting the challenges of our age
scope	structure: information as thing	process: information as event	"structuration process": both information structures and their genesis
tools	reduction: materialist monism, science culture	projection, disjunction: idealism, huma-nities culture	integration: a systematic combination of methods crossing disciplines

1.3.1 *"Normal science" information studies aims, scope, and tools*

On each science level the current paradigm in information studies faces a pluralism that extends to a cleft between divergent tendencies. It is a cleft

(1) between technocratic and ivory tower perspectives on information at the aims level,

(2) between reifying and deconstructive perspectives on information at the scope level, and

(3) between reductionistic and projectivistic as well as disjunctivistic perspectives on information at the tools level.

We are dealing with tendencies rather than clear-cut boxes into which to put existing information views. This exposed the plurality of currently held views to the tensions listed here; only in exceptional cases views be fixed to one or the other categorical "ism".

1.3.1.1 *Technocracy versus ivory tower*

First issue: first line of Table 1.2, first two columns.

There is a traditional distinction between applied sciences and basic sciences which – in a widely shared belief – is becoming increasingly blurred. The image of an engineer employed in a private lab and taking orders from his/her employer versus an academic merely satisfying his/her curiosity is old-fashioned and outdated. It is true that scientists enjoy a certain freedom of research within given financial, policy and other constraints. This, however, reflects the fact that research and development, starting in the late 1970s, have been streamlined world-wide according to neoliberal economic policies of liberalisation, privatisation, and deregulation. It is less a reflection of the general statement that science at any given time is part of society and thus responsive, be it directly or indirectly, to historically developing societal needs. This helps explain why, in developed/rich countries, many disciplines, especially in the humanities, are publicly stigmatised as being beautiful but useless and suffer cuts and total suspensions. Short-sighted economic interests have taken command in scientific affairs. Thus it still makes sense to distinguish between a business-driven development of science and technology and *l'art pour l'art* activities.

Accordingly, information concepts might either provide some foundation for ICT applications or be far detached from real-world problems.

Given the confines of economic profitability and competitiveness, the credo of technocracy is in force. Its credo: "realise everything that is feasible". This falsely presupposes that everything feasible (again, assuming it is economically reasonable) is also desirable. Accordingly, a reflective, theoretical, deliberation of norms, values, morals would be superfluous or, at best, replaced by *a posteriori*, empirical inquiries about the acceptance of technology by users. This, however, detracts from considering more fundamental problems than those of profitability. The historically most striking example for an information concept in line with this tendency is the one developed in the context of a purely technological problem (arising from military concerns) in the Bell Laboratories after World War II: Shannon's channel concept.

Conceptualisations of information devoid of considerations about potential far-reaching impacts on society are prone to being subjected to dominating economic, political, military interests. This may ultimately also be true of conceptualisations not intended to serve any specific purpose but kept within the walls of the ivory tower. Although such cases resist subsumption under instrumental reason, they are nonetheless produced under determinate historical circumstances that may unwillingly instil certain inherent values. Anything imaginable may be influenced by the state-of-the-art of already produced imaginations. Complete refusal of applicability is unreasonable.

1.3.1.2 *Reification versus deconstruction*

Second issue: second line in Table 1.2, columns one and two.

Objects of theorising and empirical investigation might be categorised as being a structure or being a process. Both categories can be viewed as being mutually exclusive and, together, exhausting the possible multitude.

Accordingly, information may be viewed as having the quality of a thing or of an event that occurs.

Screening current information concepts along this line reveals two such clusters of common perspectives.

(1) The first cluster is composed of two sub-clusters.

(a) A first sub-cluster of information concepts/theories consists of those that regard information as a given. This has been termed "potential" information or "structural" information. The structural sciences deal with that topic. They hold that matter is always in a certain shape, *gestalt*, or form, and this form is information (Bernd-Olaf Küppers is a prominent advocate of this position, which is espoused with the notion of "Strukturwissenschaften" – "Structural Sciences" – introduced in the 1970s by Carl Friedrich von Weizsäcker, see [Küppers 2000]). This form is something fixed and is equivalent to a thing.

(b) The second sub-cluster focuses on the transmission aspect. From that angle, information lies not in the structure but in that which is transmitted from a sender to a receiver via a channel (and might then be disturbed by noise). That is the classical view introduced by Shannon and is the mother of all communication models [Shannon 1948]. The transmission view of information does not see information as an event. Information is that which is carried by the signals through the channel. That which flows or floats here is sometimes referred to as "free" information. This "free" information is something fixed too, a thing that is carried.

Both sub-clusters hypostatise the phenomenon of information, they ontologise it in a specific way, they reify it. In the context of social sciences, reification means

that a relation between people takes on the character of a thing and thus acquires 'phantom objectivity', an autonomy that seems so strictly rational and all-embracing as to conceal every trace of its fundamental nature: the relation between people

– as Georg Lukács already defined it in 1923 [1972, 83].

Regarding information, this means that the social appearance of information is taken as a point of departure for making it a thing, thereby hiding its origin in social relations and dismantling its social character.

(2) Cluster number two – based on the sender–receiver–model, but extending and transcending it – focuses on what happens with, and within, the receiver. Information is ultimately not that which is transmitted – pursuant to this perspective – but that which is processed by the receiver or, more precisely, which is produced by the receiver. The receiver, by processes of decoding, attaches a meaning to the message and thereby produces "actual" information. This is the leitmotif of all developments in communication studies, in particular cultural studies, which strive to complement or depart from the channel model. The German sociologist Luhmann

deconstructed the notion of information in his social systems theory [1981, 1984]. According to him, information is by no means something lying around in the environment and waiting to be picked up. Accordingly, information cannot be transmitted either. Information is an event that occurs when expectations are frustrated and as a result of that difference another difference is triggered. This information, being an event, is hard to get a hold on. The Austrian computer pioneer Heinz Zemanek, who was involved in introducing informatics – which he did not want to call "computer science" – at the tertiary education level in Austria, insisted on the social nature of information. This makes it impossible to quantify or measure [1988].

Thus, information melts away as something fuzzy, intangible and inconceivable.

1.3.1.3 *Reductionism versus projectivism and disjunctivism*

Issue 3: third line of Table 1.2, first and second columns.

Reductionism, projectivism, and disjunctivism are different ways of conceiving the complexity of objects being investigated. They manifest themselves in the methodology of information studies.

Two different levels can be identified: (1) the first is related to the philosophical foundation of the methodology used, (2) the second is related to the disciplinary foundation of the methodology used.

(1) The philosophical considerations underlying the methodology seek the essence of information, the nature of information, the substance which makes it up. The question of what information is must be answered in relation to the essence, nature, and substance of matter.

(a) The first answer is that information is a substance equivalent to matter. This substance is conceived of as something material, making information something material. Such an answer is material(istic) monism: everything is like matter and the same holds true for information. This is materialism. Materialism in that sense is reductionistic. Under the premise that information is more complex than matter, information is reduced to matter.

(b) A second answer states that this substance is immaterial, making information something immaterial. This answer is immaterial (ideal, idealistic, ideational, informational) monism, idealism: matter is also like mind (information). Varieties include Platonism and Radical Constructivism. This, under the same premise as before, is projectivistic because information is projected onto matter.

(c) Another answer is that matter and information do not share the substance: they are essentially different in nature. Matter is material and information is not: this is the answer of dualism. Such an answer, again under the same premise as before, is disjunctivistic; it disjoins information from matter. This gives rise to yet another question: are these two substances inert and non-reactive to each other or do they interact and, if so, how then can one side of the duality affect the other side? How is it possible that matter influences mind (information)? How is it possible that mind (information) affects matter? Efforts in this direction include the Cartesian tradition (note that Descartes contended that the interaction between *res cogitans* and *res extensa* takes place in the pineal gland) and, more recently, John Eccles, who together with Karl Raimund Popper identified the synapses as the location where mind and brain interact [1977].

(2) The branch-specific considerations underlying the methodology relate to the gap between the natural and the engineering sciences (including formal sciences) on the one hand and the arts and humanities (including the social sciences) on the other hand. This dates back to the 17th century and to philosophers such as, again, René Descartes. The gap between the two branches in science reached its peak in the late 19th century with the works of Neo-Kantian philosophers, scientists, and literary intellectuals such as Wilhelm Windelband and Heinrich Rickert. Wilhelm Windelband [1894], for example, introduced the disjunction between "nomothetic" (meaning: positing laws) and "ideographic" (meaning: describing events): these would remain, existing alongside one another, as the final, incommensurable forms of our notions about

the world. Today this cleft is known as C. P. Snow's dilemma of the two cultures, which he bemoaned in 1959 and 1963 [1998]. Although John Brockman, an author from the USA, foresees the emergence of a third culture "founded on the realization of the import of complexity, of evolution" [1995, 20-21], this is by no means mainstream. The cleft is characterised by the preponderance of either the analytical method or phenomenological and hermeneutic methods, of so-called third-person science versus first-person science, of an externalist versus an internalist view: each characterises one or the other tendency. The same holds true for the information concepts.

(a) The first approach is, methodologically, reductionistic. It reduces different qualities of the phenomena under investigation to one and the same quality, typically the quality which is the simplest.

(b) The second approach is characterised by a humanistic rationality which is ignorant of the field of science and technology. Methodologically, there are two possibilities.

- The first projects a particular quality (typically the most complex one) onto phenomena which lack this quality, and then purports to discover them there. Properties of information in nonhuman domains are usually extrapolated from properties of information in the human domain (anthropo(socio)morphism).

- The second gives up the attempt at a subsuming, though unifying, solution and argues instead in favour of a lack of comparability of the given phenomena in nature and society. In this dichotomising, disjunctive, view, information is exclusively ascribed to the human domain. Going even further, it may exclusively be ascribed to particular incidences within the human domain.

1.3.2 *A Science of Information paradigm*

Accordingly, information studies appear to have already passed the normal science phase and to have entered a critical phase. This is because, at each level – be it aims, scope or tools – a possible position is countered by a possible counter position. Precisely these discrepancies form an obstacle to unification. Progress in the direction of a Science of Information implies efforts to overcome these divides.

Those efforts should help take up what is important while discarding the one-sidedness of any particular theory in the field. The search for integration potentials between different theories should do justice to their input and at the same time reflect caveats concerning possible blind spots.

1.3.2.1 *Ensuring futurability*

First issue: line 1, column 3 of Table 1.2.

Concerning the opposition of applied science and basic science, Pasteur's Quadrant offers a solution. Donald Stokes [1997, 12] produced a 4 times 4 matrix with the "Quest for fundamental understanding?" as one dimension and "Consideration of use?" as the second one, with yes and no answers for each. For him, Thomas Alva Edison is the role model for pure applied research, Niels Bohr the model for pure basic research: Louis Pasteur, however, is paradigmatic for a new way of doing science, namely "use-inspired basic research".

A Science of Information certainly belongs to Pasteur's Quadrant and must be open to practical aspects if it is to successfully support the search for a fundamental understanding of information. Success, however, must not be understood in a restricted economic, political, military or technological sense. What is needed is openness in considering the great challenges facing humanity today and prioritising values accordingly. The *raison d'etre* of a Science of Information is to provide society with a means of enhancing its problem-solving capacity *vis-à-vis* these challenges, to provide society with a future, to make it futurable.

The underlying problems consist of frictions in the functioning of society, environment and technology. Problem-solving activities seek information applications that reduce the frictions. Accordingly, a Science of Information would be the safeguard against the loss of control, it would guarantee the stabilisation of development and the maintenance of society (Figure 1.2). In this sense, it is destined to become a, if not the, science for the information society.

The vision of a "good society" must serve as a point of departure [Hofkirchner 2011a]. In this sense, a Science of Information is normative. Technological applications are to be closely reviewed, and the question is: are they apt to serve the purpose of a good society?

The design process must start by identifying a societal problem and then proceed with the search for appropriate applications (and not vice versa, as is done under technocratic premises).

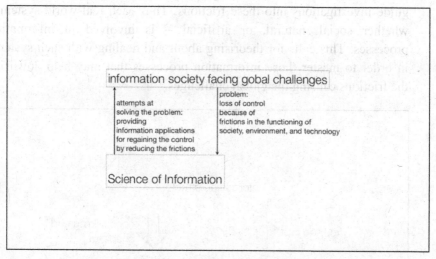

Figure 1.2. The role of Science of Information in society.

In that context, a Science of Information implies a transcendence from the scientists to the stakeholders and those affected by the results of research. It implies a transformation into a new science that is human-centred, democratic, participatory, such as Helga Nowotny's "Mode-2" science [Gibbons and Nowotny 2001].

1.3.2.2 *Catching the ephemeral*

Second issue: line 2, column 3 of Table 1.2.

A future Science of Information is not merely normative. It also does justice to the factual. In evaluating the potential that selects the desired, it also accounts for the potential that is given with the actual. Accordingly, it is clear that the domain of this Science of Information encompasses anything that promotes or stalls a good society. It identifies the frictions that cause malfunctions in society, environment and technology. And it seeks potentials of ephemeralisation.

Importantly, society has used the informational revolution to, theoretically, boost its potential to reduce frictions wherever they appear. This reflects the ongoing information processes between all parts of society, environment and technology. What is missing is a science to guide investigations into these frictions. Thus each real-world system – whether social, natural, or artificial – is involved in information processes. This calls for theorising about and dealing with such systems in order to master those information processes that may help downsize the frictions causing the global challenges.

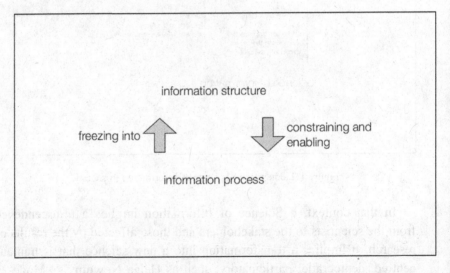

Figure 1.3. Information structuration process.

To understand these phenomena, information must be understood as both structure and process, i.e. as a "structuration process"[f] in which processes produce structures that, in turn, structure the processes, that is,

---

[f] I use here the term coined by Anthony Giddens with regard to societies [1984].

function as both constraints and enablers for the continuation of the processes. In fact, frictions in the information processes depend on the very relationship of constraining and enabling (Figure 1.3).

The notion of "structuration process" would revive the ancient meaning of *informatio*, albeit in a new context.

Information can be viewed as something overarching the whole bandwidth of different and diverse structuration processes in our universe and manifesting itself in a variety of phenomena.

1.3.2.3 *Taking the blind men's perspective*

Issue 3: line 3, column 2, Table 1.2.

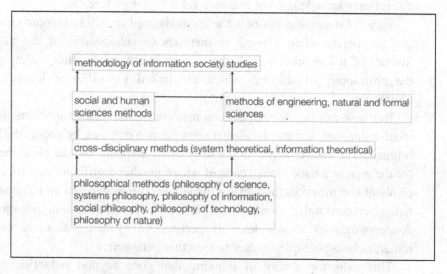

Figure 1.4. Interdependence of Science of Information subdomain methods.

The investigation must comprise the wide range of matter pointed out above. Thus, a Science of Information cannot, with reference to the tools, afford to neglect any potentially fruitful and elucidating methodological approaches. Likewise it must not fail in putting the puzzle of findings together and in synthesising the manifold analyses. This would help transcend the borders of disciplines and strive for the unity of science based on a unifying approach, without subjecting any thinking to uniformity.

Transdisciplinarity defines concepts that go beyond the meaning of multi- and even interdisciplinarity. Multidisciplinarity means the unrelated coexistence of monodisciplinary accounts, and interdisciplinarity means the casual establishment of relations between monodisciplines without feedback loops that lastingly impact their repertoire of methods and concepts. Transdisciplinarity, in contrast, comes into play when each discipline is engaged in the collaborative undertaking of constructing a common foundation of methods and concepts, of which its own methods and concepts can be understood as a type of instantiation. Transdisciplinarity does not mean abolishing disciplinary knowledge, but grasping for a bigger picture.

Figure 1.4 summarises how the methods used in different approaches may cooperate when viewed as methods of subdomains of the new Science of Information – from the science of the information society to the philosophy of science, which are linked via different levels of abstraction.

But how can the divide between materialist and idealist monism and idealist dualism, and the divide between the two cultures, be successfully bridged? The answer is like in the story of the elephant and the blind men (or the men in a dark room), each of whom touches a different part of the elephant and mistakes the part for the whole [Blind Men and an Elephant n.d.]. Accordingly, none of the various existing information concepts/theories should take its perspective as being absolute but, rather, as being complementary to the other perspectives.

This calls for a way of thinking that goes beyond reductionism, projectivism, disjunctivism. What is needed is *"unitas multiplex"*, as the French philosopher and sociologist Edgar Morin calls it [1999, 25, 19]:

> It means understanding disjunctive, reductive thought by exercising thought that distinguishes and connects. It does not mean giving up knowledge of the parts for knowledge of the whole, or giving up analysis for synthesis, it means conjugating them. This is the challenge of complexity which ineluctably confronts us as our planetary era advances and evolves.

This is the integrative way of thinking that a Science of Information will need to incorporate.

Part 2

Steps to a
Unified Theory of Information

Chapter 2

A New *Weltanschauung*

The Planetary Era demands that we situate everything in the planetary context. Knowledge of the world as world has become an intellectual as well as a vital necessity. It is the universal problem of every citizen: how to gain access to global information, and how to acquire the possibility of linking together and organizing it. To do so, and thereby recognize, acknowledge, and know the problems of the world, we need a reform in thinking.

– Edgar Morin: Homeland Earth, A Manifesto for the New Millennium, 1999 –

Evidence ranging from physical cosmology through the biological to the human and social sciences indicates that events and entities in this universe are dynamic: systems evolve in space and time [...] open-system evolution is toward higher specific negentropy in the evolving open systems, achieved through greater structural and functional complexity and producing a higher throughput of free-energy flux density and more extensive interaction between systems and environments.

– Ervin Laszlo: Where No System is Entirely Closed, 1998 –

The contention here is that a paradigm shift is underway that might be more secular than any other before. It goes far beyond any single discipline. It affects all science and everyday thinking as well.

The fact is that we are facing a point in the development of humanity at which, for the first time in history, mankind has acquired the means for committing omnicide, i.e. the suicide of *Homo sapiens sapiens* and the possible "lunification"[a] of earth. This situation is due to the success of the scientific-technological progress made since the instauration of

---

[a] This term was coined by the Austrian philosopher Leo Gabriel († 1987) when he was president of the Vienna *Universitätszentrum für Friedensforschung*, which helped facilitate the East-West dialogue during Cold War.

Francis Bacon's programme of empirical science. Now that the evil side of its implementation has come evident in the global *problematique*, time for reflection has come. The programme needs to be reworked, not so Bacon's ideals [Schaefer 1993, Hofkirchner 2003]. The paradigm shift addressed here is about the self-reflection of human civilisation whose course needs to be changed in order to persist. This self-reflection requires cutting across the disciplines throughout science, including philosophy, and extending it to everyday thinking.

Thus it might be called a new *weltanschauung*. A *weltanschauung* differs from, but includes, a view of the world in that

(1) it is not conceived from the perspective of a single discipline but integrates different scientific viewpoints;
(2) it is not merely a description but also normative;
(3) it is not confined to science and philosophy but extends to common sense.

Philosophy might be viewed as the scientific kernel of *weltanschauung*.

2.1 A New Philosophy

Philosophy refers to the most fundamental reflection about humans and their position in the world.

At least three fundamental questions that have constituted philosophy from the onset, although they have been formulated in different ways.

(1) One question deals with values, norms, imperatives, guidelines for acting.
(2) Another question deals with the world as it is, its properties, be it with or without us.
(3) A third question deals with our ability to produce knowledge.

The first question makes up the domain of ethics, aesthetics and axiology, all of which address practice, the second makes up the domain of ontology, which is about reality, and the third makes up the domain of epistemology including the method(ology) of inquiry.

These three domains may be tackled either as separate fields of philosophy, or as networked or even nested.

Historically speaking, awareness of the interdependencies of the fields has been growing. The philosophical stance developed here is termed Praxio-Onto-Epistemology (POE). Its main feature is the integration of the three domains by employing a classification of ways of thinking.

2.1.1 *A new way of thinking*

What is a way of thinking? It describes the way how identity and difference are thought to relate to each other. Relating identity and difference is probably the most basic function of thinking. It appears in praxiological, ontological and epistemological contexts. Accordingly, practical problems that come to thought, entities that are investigated, phenomena that have to be cognised, may be identical in certain respects but may differ from each other in other respects.

Identity and difference can be viewed from the point of complexity. Then that what differs is more complex than that from which it differs. This raises the question as to how the simple relates to the complex, i.e. how less complex problems or objects or phenomena relate to more complex ones (Table 2.1).

Table 2.1. The four ways of thinking.

ways of thinking	relationship between lower and higher complexity	relationship between identity and difference; unification or diversification
reductionism	reduces higher complexity to lower complexity	
projectivism	projects higher complexity onto lower complexity	identity without difference; unification
disjunctivism	disjoins higher complexity from lower complexity	difference without identity; diversification
integrativism	integrates lower with higher complexity and differentiates between them	unity of identity and difference; integration

(1) The first way of thinking, in terms of ideal types, establishes identity by eliminating the difference to the benefit of the less complex side of the difference and at the cost of the more complex side. It reduces "higher complexity" to "lower complexity"; this is known as reductionism. Reductionism is still the main stream of natural science.

(2) The counterpart of the reductive way of thinking is what might be termed "projective". Projective thinking also establishes identity by eliminating the difference, albeit to the benefit of the more complex side of the difference and at the cost of the less complex side; it takes the "higher" level of complexity as its point of departure and extrapolates or projects from there to the "lower" level of complexity. It overestimates the role of the whole and diminishes the role of the parts. This is a trait of the humanities. Both the reductive and the projective way of thinking yield unity without diversity.

(3) The third way is opposed to both the others in that it eliminates identity by establishing the difference for the sake of each manifestation of complexity in its own right. It abandons all relationships between all of them by treating them as disjunctive; it dissociates one from the other, dichotomises and yields dualism (or pluralism) in the sense of diversity without unity. It is here termed disjunctivism. The often bemoaned cleft between the so-called two cultures of hard science and soft science (humanities) is the most striking example for this way of thinking. In fact, this is a description of the state of scientific endeavour as a multiplicity of monodisciplinary approaches that are alien and deaf toward each other.

(4) The fourth way of thinking – integrativism – is again opposed to all the others, i.e. to reductionism, projectivism and disjunctivism. It establishes identity as well as difference, favouring neither of the manifestations of complexity. It establishes identity in line with the difference. It integrates both sides of the difference (yielding unity) and it differentiates identity (yielding diversity). This way of thinking is based upon integration and differentiation. It is opposed to both dissociation and unification and yields unity and diversity in

one. It integrates "lower" and "higher complexity" by establishing a dialectic relationship between them[b].

2.1.1.1 *Reductionism*

The relationship between practical problems, entities and phenomena of different complexity might be visualised using set-theoretical considerations as follows: be A the simple and B the complex. Then A is the basic set, B a subset (and B' the complementary set) (Figure 2.1).

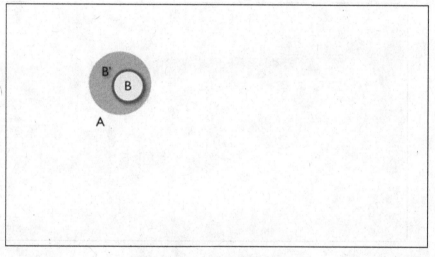

Figure 2.1. Identity and Difference.

This means: B is differentiated, given the basis A; B is different, it differs from all A that is not B (which is B'); A represents the identity of both sides of the difference (B and B').

---

[b] A dialectic relationship is said to exist, if the following criteria come true: firstly, both sides of the relation are opposed to each other; secondly, they depend on each other; thirdly, they are asymmetrical in that neither side can be replaced with the other without simultaneously replacing the mode of relationship. Master-and-slave or mother-and-daughter are examples for dialectic relationships.

Reduction is a logical operation that eliminates the border between B and B' such that B becomes A and merely A (Figure 2.2). Reductionism is the habit of taking for granted that every B is simply A.

A praxic example: let A be the knowledge about life in prehuman nature and B the values for human conduct. Then reductionism would see the values as consequences of how nature is behaving. So, for instance, the "struggle for life" in nature is seen as the justification of the "struggle for life" in society.

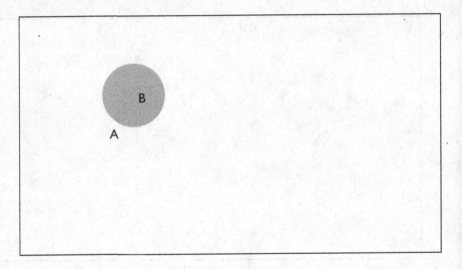

Fig. 2.2. Reduction turns B into A.

An ontic example: let A be living nature and B humans. Then reductionism conjures away the difference humans make in living nature. Accordingly, ethology looks upon humans as naked apes.

An epistemic example: let A be biology and B sociology. Then reductionism tries to draw social conclusions from biological premises. "Sociobiology", for example, contends that biology is the discipline that explains sociology[c].

---

[c] For that reason, Sociobiology should, more precisely, be termed "biosociology". "Sociobiology" would rather mean biology that is shaped by sociology.

2.1.1.2 *Projectivism*

Projection is a logical operation that eliminates the border between B and B' such that A becomes B and fully-fledged B (Figure 2.3). Projectivism is the habit of taking for granted that every A is completely B.

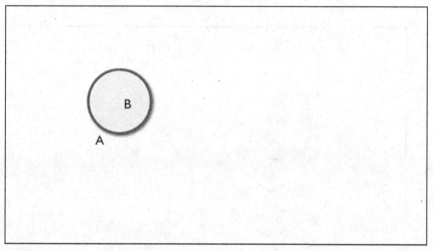

Fig. 2.3. Projection turns A into B.

The praxic example: according to projectivism, values of human conduct are valid for nature as well. The "struggle for life", originally found in society, is expanded to cover prehuman nature too.

The ontic example: according to projectivism, living nature does not differ from humans. Anthropomorphistic fallacies extrapolate human traits to nonhuman species and organisms, for example the "mind".

The epistemic example: according to projectivism, biological conclusions shall be derived from sociological premises. Ant colonies and bee hives are likened to human states with queens, workers etc.[d]

Note that both reductionism and projectivism arrive at the equation A = B but that in the first case it means A = A and in the second one B = B.

---

[d] The term "Sociobiology" would fit here.

2.1.1.3 *Disjunctivism*

Disjunction is a logical operation that removes the common ground of B and B' such that A and B become solely A or B (Figure 2.4). Disjunctivism is the habit of taking for granted that A and B are disjunct.

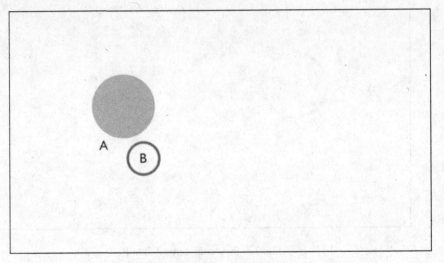

Fig. 2.4. Disjunction separates A from B.

The praxic example: the realm of behaviour in nature and the realm of human conduct have nothing in common: there is no connection whatsoever between natural facts and human values.

The ontic example: living nature is extrahuman: animals don't have souls.

The epistemic example: biology and sociology in no way depend on one another: propositions about the social life of living systems and propositions about humans in society use incompatible languages and there can be no translation.

2.1.1.4 *Integrativism*

Integration (with differentiation) is a logical operation that sees B and B' integrated into A and, at the same time, A differentiated into B and B' (see again Figure 2.1). B is the result of a differentiation in A.

Integrativism is the habit of taking for granted that A is the general ground for B and B is a specification of A.

The praxic example: human values of conduct, be it competition-oriented or cooperation-oriented, cannot be derived from nature, but nature provides biotic foundations for either conduct.

The ontic example: humans are a product of the evolution of nature. In the course of speciation, humans made such a difference that the new features of mind, language and work began to outbalance the dynamics of biological evolution and even change their physiological nature of humans; nonetheless, humans will never be cut free from surrounding nature or from their inner nature.

The epistemic example: sociology describes its part of reality with particular concepts and theories which relate to the concepts/theories of biology in that the latter describe spaces of possibilities for the realisation of social subject matters. This enables the quest for a common language.

Integrativism is itself a type of dialectic sublation of unification and dissociation, of reductionism and projectivism and disjunctivism. A dialectic sublation eliminates the dominant role of the preceding quality rather than the quality itself. This quality is kept, i.e. continued, but it is continued under the dominance of a new quality and is therefore – as Hegel phrased it – lifted onto a next level. All three meanings of sublation are valid for the integrativist thinking *vis-à-vis* the fallacious ways of thinking. Reductionism, projectivism and disjunctivism are not totally negated but taken *cum grano salis*. Each of them has an aspect of overexaggeration that has to be eliminated but, by the same token, has an aspect that is correct once the onesidedness is removed. The novel integrative view does justice to these aspects; this approach establishes unity among the diverse conflicting views.

Hereby, the principle of unity-through-diversity is established. The nonintegrativist ways of thinking yield either unity without diversity (in the first and second case of reduction and projection) or diversity without unity (in the third case of disjunction). Integration, as defined here, yields unity in line with diversity, unity in diversity, but also diversity in unity. Diversity is considered to be a necessary condition for unity. Thus it is called "unity-through-diversity" (Figure 2.5).

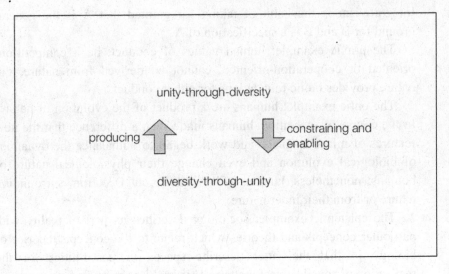

Fig. 2.5. Unity-Through-Diversity.

The following definition can be given:

Integrativism. *Integrativism is that way of thinking that applies the Principle of Unity-Through-Diversity.*

> The **Principle of Unity-Through-Diversity** states: Diversity is a necessary, though not sufficient, condition for unity and, in turn, unity is a necessary, though not sufficient, condition for diversity.

Diversity and unity condition each other. Diversity can produce unity (in which case we refer to unity-through-diversity), but need not do so. Unity can enable diversity (in which case we refer to diversity-through-unity), but it can constrain diversity to uniformity (eliminating unity-through-diversity).

In summary, integrativism opposes reductionism and projectivism as well as disjunctivism, but considers reduction, projection or disjunction as justified within certain boundaries when taking into account the legitimate claims of each other.

Integrativism is one step on the way to a Unified Theory of Information.

2.1.2 *A new concept of practice–reality–method*

Integrative thinking serves here as guidance to interrelate the three domains of philosophy that deal with practice, with reality, and with method.

After times of prevailing realistic stances in ontology, varieties of radical constructivism arising in the second half of the 20th century have emphasised on the interdependence of epistemological and ontological questions (though some of them tended to end up in mere solipsism). According to them "the world as it is" is difficult to approach because our assumptions of how the world is turn out to rely on specific methods of human cognition. A moderate stance trying to reconcile realism with constructivism is onto-epistemology as coined by Hans Jörg Sandkühler [1990, 1991, 34-37, 353-369] and shaped by Rainer Zimmermann [2002, 147-167]. The position developed in the present contribution takes up this line of thought and strives to complement the interrelationship of ontology and epistemology by establishing the relation to ethics, aesthetics and axiology. The latter complex is proposed to be included in so-called "praxiology".

Praxiology is meant as the philosophical theory of praxis. Praxis refers to society and its human actors. It can be defined as the process of co-action (of human co-actors) upon reality which, in turn, can be defined as the field of interaction (of human actors with the environment) that is mediated by method; the latter, finally, can be defined as the way of action (of those human actors). In terms of subjects and objects, praxis is the totality of the human subject-object-dialectic, reality is what is, so to say, "objecting" to becoming "subject" to humans, and method is the subjective way of making objects "subject" to humans. Clearly, based on that order of definitions, praxis builds upon reality and reality builds upon method, be it material or ideational. Hence, if ontology is meant as the philosophical theory of reality and epistemology as the philosophical theory of the method of inquiry, praxiology is based upon ontology and ontology on epistemology. This relationship of encapsulation of domains of philosophical disciplines, and of the philosophical disciplines themselves, does not model a one-way process of causal influence or linear inferences. Rather, both bottom-up as well as top-down processes have to be recognised. There is relative autonomy of each of the domains (praxis may shape reality but reality provides the scope of possible practices; reality may shape method but method provides the scope of possible realities) and of each of the disciplines (praxiology does not fully determine ontology and ontology does not fully determine epistemology, and *vice versa*).

The rationale for defining subject matters and respective theories in such a concatenated way is that it enables providing an appropriate sketch of the following relations: certain interests (that reflect certain practices) define the sphere of intervention (that is made up of objects in which subjects are interested and is characterised by a boundary beyond which there are no real objects because there is no subject interested in them). Equally, certain spheres of intervention (that reflect certain realities) define the scope of instruments (that is made up of means which are useful for intervention and is characterised by a boundary beyond which there are no real means because they do not fit the object). Finally, certain instruments (that reflect certain methods) can help construct a certain sphere of intervention only by excluding different realities, and certain spheres of intervention can meet a certain group of interests only by excluding different interests.

Thus, explicitly taking a human stance, we can reformulate the fundamental questions of philosophy by starting with the praxiological question and subsequently introducing the ontological question and the epistemological question, each one being the presupposition for the question before (Figure 2.6):

(1) How should humans act, or better, what should the world be like?

(2) How can humans intervene in the world, or better, how can humans make the world be as it should be?

(3) How can humans comprehend the world, or better, how can humans know how to make the world be as it should be?

Having guidelines for action presupposes having ideas about where human actors start from, and having ideas about where human actors start from presupposes having tools to recognise the starting point. If humans want to succeed in changing the world, they need to comprehend the relations that allow achieving the goals they have set. And in order to gain this knowledge, they must apply all potentially worthwhile means.

Accordingly, the praxio-onto-epistemological standpoint is one in which praxiology does matter: ontic propositions bear the stamp of practical instructions and pass this stamp over to epistemic methods.

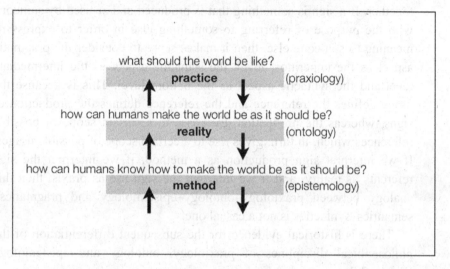

Fig. 2.6. The order of the fundamental questions of philosophy in Praxio-Onto-Epistemology.

Clearly, by defining praxis as the subject matter of praxiology, this term denotes the philosophical theory of human actions in regard not only to their efficiency, effectiveness and efficacy, but also according to moral value and beauty – see different views in [Mises 1999], [Kotarbinski 1965], [Bunge 1999], [Collen 2003].

This approach to encapsulating praxiology, ontology and epistemology may be regarded as a generalisation of what action theory points at when establishing a relationship between ends, ways and means. Ends refer to practical interests, ways to ontic interventions, and means to epistemic instruments.

Furthermore, the manner in which the nestedness of praxiology, ontology, and epistemology is conceived resembles the manner in which semiotics can be structured. Semiotics is about signs and can be divided into the subdisciplines pragmatics, semantics and syntactics. According to the perspective advocated here, semantics is a subset of pragmatics, and syntactics is a subset of semantics. Pragmatics deals with the use of signs, semantics is about the signs' relation to their referents, and syntactics refer to signs in themselves and how their components fit together. If a sign is something that is produced and/or used by someone with the purpose of referring to something else in order to express its meaning to someone else, then it makes sense to consider the pragmatic aspect as the uppermost level, the semantic aspect as the intermediate level and the syntactic aspect as the bottom level. This is because the usage defines the reference and the reference defines the production of signs, whereas the production gives rise to a certain scope of possible references which, in turn, gives rise to a certain scope of possible usages. If we interpret sign production as a method, if we interpret the sign referents as reality, and if we interpret the sign use as praxis, then this analogy between praxiology–ontology–epistemology and pragmatics–semantics–syntactics is not a casual one.

There is historical evidence for the subsequent differentiation of the philosophical disciplines of praxiology, ontology and epistemology (Figure 2.7).

(1) A first stage of development is mythology, which in ancient societies was a reflex of human praxis as an undifferentiated whole.

(2) Another stage of development is characterised by the advent of the philosophy of nature as developed in Greek civilisation and contemplating the object of practices.

(3) A third stage of development, finally, is introduced by specialised philosophy of science considerations that initiated the self-reflection of human inquiry in the 20th century.

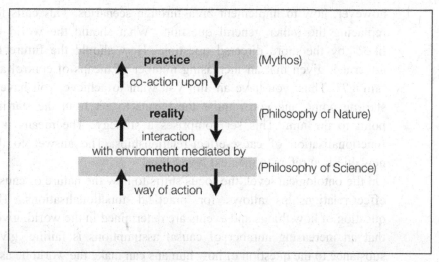

Fig. 2.7. The order of levels in Praxio-Onto-Epistemology in a historical perspective.

Praxio-Onto-Epistemology is a response to the current developmental requirements of humanity.

(1) On the praxiological level, the survival of civilisation necessitates establishing a new world order; this is tantamount to achieving a unity of the social world through a diversity of the social players populating it.

(2) On the ontological level, the unity of the social world we live in presupposes the unity of the physical world through a diversity of physical things and events that constitute the physical world.

(3) On the epistemological level, the unity of the physical world must be reflected by the unity of the world of knowledge through a diversity of disciplinary approaches.

Thus, due to the specific point of history that humankind has reached, on each level, a particular question arises that instantiates the fundamental philosophical questions listed in the section above.

(1) On the praxiological level, the focus is on how to regain control over a clearly uncontrolled development of human civilisation that might lead to catastrophe. The ideal of how the social world should look like can, in principle, be substantiated by choosing between scenarios of a breakthrough and scenarios of a breakdown. The question is,

however, how to implement breakthrough scenarios. This calls for replacing the rather general question "What should the world be like?" by the more directed question "How should the future be governed, given that an increasing number of means of control are failing?". Thus, you have an aim you want to achieve, you have a starting point, and you require the means to get from the starting point to the aim. This set comprises a strategy. The means is a functionalisation of cause-effect-relationships. The answer to the new question refers to the next lower level.

(2) On the ontological level, the focus shifts to how the nature of cause-effect-relationships allows for practical functionalisations. The question of how things and events are determined in the world, given that an increasing number of causal assumptions is failing, gives substance to the question of how humans can make the world be as it should be. Cause-and-effect relationships are assumed to work if *explanans-explanandum* reasoning works. This raises the question for determinism and, finally, another question on the lowermost level.

(3) On the epistemological level, the question is "How can explanations help understand the world, given that an increasing number of conjectured explanations are failing?". This specifies the issue of how humans can know how to make the world look like it should be.

The following reformulations help further focus the questions. (1) The basic alternative praxiology faces today seems to be: "controllability vs. uncontrollability"; (2) the basic alternative of ontology today seems to be: "determinacy vs. indeterminacy"; (3) the basic alternative of epistemology today seems to be: "explainability vs. unexplainability" (Table 2.2).

However, while traditional approaches are apparently stuck in these alternatives, POE offers a response going beyond these alternatives. Based upon the new way of thinking (integrativism), POE attempts to answer these questions by abiding to the integrative Principle of Unity-Through-Diversity. This distinguishes it from traditional philosophical approaches.

Table 2.2. Three basic questions of Praxio-Onto-Epistemology.

praxiology	ontology	epistemology
controllability vs. uncontrollability	determinacy vs. indeterminacy	explainability vs. unexplainability
the feasible and the desirable	cause and effect (old and new, parts and whole)	essence and phenomena

Thus, cross-tabling the four ways of thinking and the three basic questions of POE yields a twelve-box table, in which each line signifies one consistent philosophical answer (Table 2.3).

Table 2.3. Philosophical trains of thought concerning the basic questions of Praxio-Onto-Epistemology, classified according to the four ways of thinking.

	praxiology		ontology		epistemology	
reductionism		practicism		preformationism, atomism		scientism
projectivism	activism	utopianism, romanticism	determinism	teleologism, holism	rationalism	anthroposociomorphism
disjunctivism	inactivism		indeterminsm		irrationalism	
integrativism	deliberate activism		less-than-strict determinism		reflexive rationalism	

The first line, composed of practicism, preformationism and atomism, and scientism reflects – in an ideal-typical way – the mainstream of Modernity as modernist, mechanicist-materialistic, and naturalistic (including trains of technicalisation and mathematical and logical formalisation), respectively.

The second line, with utopianism and romanticism, with teleologism and holism, and with anthropomorphism, is the twin of the mainstream of Modernity – an opposite that nevertheless belongs to the same system[e]. Fundamentalist anti-modernism, idealistic mysticism, and anthroposociomorphistic culturalism, respectively, are characteristic of that counterpart.

Both lines of thought have something in common, giving them a "classical"[f] quality. They share an activistic, deterministic and rationalistic tendency.

The third line is composed of post-modern anti-modernism, dualistic mysticism and culturalism in the sense of Snow's two-cultures. It might be called "nonclassical", insofar it is inactivistic, indeterministic and irrationalistic. The nonclassical train of thought represents an unresolved contradiction of a submerged humanistic tradition against the suppression of feeling, as outlined by Stephen Toulmin [Toulmin 1990].

The last line – the "postnonclassical" cluster – tries to do justice to both the "classical" and "nonclassical" strands while overcoming their one-sidedness. It does this by promoting the idea of the unity of practice, reality, and methods with deliberate activism, less-than-strict-determinism, and reflexive rationalism, respectively.

---

[e] Note that, shortly after the 9/11 event, French philosopher Jean Baudrillard [2002] defined the role of Al-Qaeda in a similar manner: as being the natural counterpart of the Western-dominated world system having its origin in that very system, a symptom of the evil, the terrorist twin of globalisation.

[f] I borrow the wordings "classical–nonclassical–postnonclassical" from Vjacheslav Semenovich Stepin from the Institute of Phiosophy of the Russian Academy of Science. He introduced them to the scientific community more than ten years ago, albeit in Russian (personal communication, Iryna Dobronravova). My wordings may have a somewhat different meaning.

The subsequent chapters (3 to 5) discuss the boxes of the table in detail, column by column. First, however, the next section introduces the cross-disciplinary level that fits the philosophical part of the new *weltanschauung*.

2.2 A New Cross-Disciplinary Paradigm

The new way of thinking and the new concept of practice–reality–method, that is, the new philosophy, extends to the cross-disciplinary partition of *weltanschauung* that surrounds its kernel. This purpose is to make the new paradigm complete.

We live in an age of global problems that concern the survival of humanity. It is the nature of these challenges to be complex. Accordingly, they have to be approached in a similarly complex fashion. The response to this need has created a more far-reaching paradigm shift than ever before, focussing on cross-disciplines. What is known as "sciences of complexity", "self-organisation theories", "evolutionary systems theories", are elements, if not the core, of this overall shift.

The foundations for this shift were laid when Ludwig von Bertalanffy founded "General System Theory" [Hofkirchner 2005] in conjunction with other scholars like Anatol Rapoport, Kenneth E. Boulding or Ralph W. Gerard.

Heinz von Foerster is the first who, in the late 1950s, introduced the notion of self-organising systems to the scientific community (though at that time the notion had a somewhat different and more restricted meaning). Von Foerster published in 1960 the article *On Self-Organizing Systems and Their Environments*, one of the first and most important works on this topic [Foerster 1960]. In 1961 he organised a conference on self-organising systems in Chicago and, together with George Zopf, edited the proceedings in 1962 [Foerster et al. 1962].

Today, the common understanding of "self-organisation" is a process or event in which matter displays its capability of spontaneously building up order and maintaining it. The physical basis for self-organisation was already resesarched by a team around Ilya Prigogine already in the 1940s (emergence of macroscopic structures in dissipative systems that have moved far away from thermo-dynamic or chemical equilibrium [Nicolis and Prigogine 1989]). Since the 1960s the notion of self-organisation has been applied in various disciplines, and empirical evidence has been found *en masse*. After Prigogine, Hermann Haken generalised the physics of self-organisation to his so-called Synergetics, with applications in social and economic science as well (order out of chaos, slaving principle, [Haken 1978, 1983]). Manfred Eigen [Eigen and Schuster 1979] described the emergence of living matter in a hypercycle of autocatalytic reactions. Humberto Maturana and Francisco Varela [Varela et al. 1974] put forward their idea of living systems as autopoietic ones, which can reproduce and maintain themselves. Niklas Luhmann [1984] applied autopoiesis to society by suggesting that social systems are self-reproducing ones. The examples are only the tip of the iceberg. Today, the theory of self-organisation has diffused into nearly all scientific disciplines.

More importantly, considering self-organisation enabled system theory to depart from a state in which it could deliberate only on how systems are maintained and to include system changes. At the same time, it opened the possibility for the Theory of Evolution to overcome the restrictions of simple mechanistic interpretations of the Darwinian model. This step enables the sciences to envisage a theory of open, non-linear, complex, dynamic, self-organising systems. It is the direct result of the merger of systems theory and evolutionary theory.

It is therefore apt to call this theory "Evolutionary Systems Theory" (EST) – a term coined by Ervin Laszlo [1987], Vilmos Csanyi [1989] and Susantha Goonatilake [1991], but used here to extend the meaning it had worldwide until the *Konrad Lorenz Institute for Evolution and Cognition Research* Seminar held in Vienna in 1995 [Van de Vijver et al. 1998]. EST is understood here as a theory about evolving systems and as a theory of systemic evolution. EST no longer deals merely with mechanisms, strategies and controls for achieving or maintaining homeostasis and the development of species; it also concerns the birth, growth and decline, i.e. development, of systems, from the formation of the earliest known particle, through the arrival of terrestrial life forms, to the shaping of specific social systems on the human level – it concerns the rise and fall of all real-world systems [Ebeling and Feistel 1994,

Goerner 1994, Mainzer 1994], including the cosmos itself [Layzer 1990, Smolin 1997]. The concepts of dissipative structures, synergetics, hypercycles, autopoiesis and self-referentiality are the most prominent predecessors of a theory of evolutionary systems that deals with manifold topics. These topics include the formation of structures by means of coherent behaviour of particles in thermodynamics or laser-active materials, the formation of spatial or temporal patterns in chemical reactions, the origin and development of highly complex molecules as prerequisites for the formation of biotic systems, phylogenesis (in which the adaptation of systems to their changing environments is seen as the achievement of the systems themselves rather than the achievement of the environment as implied by old evolutionary theories) and ontogenesis, the relation of matter and mind and cultural phenomena.

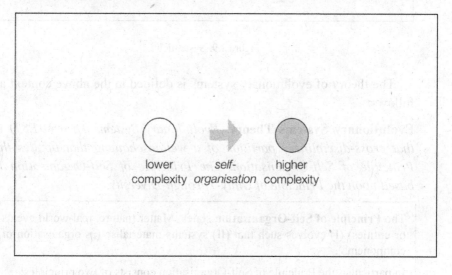

Figure 2.8. Evolvability.

EST accounts for the evolvability as well as systemicity of matter, nature or real-world events and entities. Thus, self-organisation is involved in two different fields [Hofkirchner 2000]:

(1) it conveys the transformation of systems such that a spiral of increasing complexity appears in the course of evolution [Ebeling and Feistel 1994] (Figure 2.8),

(2) and it provides the scaffolding of systems such that organisation is maintained by the components (Figure 2.9).

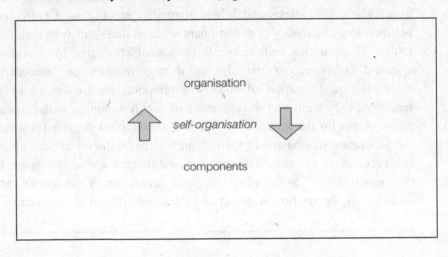

Figure 2.9. Systemicity.

The theory of evolutionary systems is defined in the above context as follows:

Evolutionary Systems Theory. *Evolutionary Systems Theory (EST) is that cross-disciplinary partition of a weltanschauung that applies the Principle of Self-Organisation. The Principle of Self-Organisation is based upon the Principle of Unity-Through-Diversity.*

The **Principle of Self-Organisation** states: Matter (nature, real-world events or entities) (I) evolves such that (II) systems materialise (as organisation of components).

In particular, the Principle of Self-Organisation consists of two principles:

(I) The **Principle of Evolvability** states with regard to diachrony: Matter is evolving from lower complexity towards higher complexity, through complexification and simplification, through differentiation and integration. Thereby "complexification" means the increase in differences, "simplification" the integration of differences; "differentiation" means the emergence of at least one difference, "integration" the relinking of both sides of a difference to what they have in common.

> (II) The **Principle of Systemicity** states with regard to synchrony: Systems are the materialisation, that is, the fall-out or the condensation, of evolution through the organisation of components.

Evolution and systems are intertwined like a river and its bed: the process crystallises in a structure and the structure channels the process.

EST comprises systems in line with complexity science considerations such that, revolving around the notion of self-organisation, it provides a transdisciplinary framework for consilience throughout science.

Accordingly, all science serves to support efforts to master the global challenges. Moreover, EST has prompted an increasing number of researchers to discovering evolutionary systems regardless of which real-world object they may be investigating. This is because providing specialised knowledge about the functioning of different self-organising systems is essential to influence such systems in such a way as to trigger the most promising development paths. Finally, in EST, diverse methodological approaches are viewed less and less as impediments that endanger the unity of science; rather, they are increasingly regarded as useful means towards the same end and as an enrichment of science, as long as the common basis of the different methods is not violated.

Chapter 3

Nudges[a]

[...] A last cluster of theories can be characterised by their radical parting with an understanding of steering that has to do with anything like goal oriented control. Instead, they view governance of sustainable development as a task of modulating open-ended developments which are driven by a complex interplay of the inner dynamics of actors, subsystems, and bio-physical processes. The theoretical background is formed by concepts such as self-organisation, complexity, or co-evolution. In this view development is contingent and future structures are emergent phenomena that cannot be influenced by single factors (like a strategy of a particular actor) but that come into existence only in the relation and interaction of many diverse factors. [...] Nevertheless, even on this yielding theoretical ground it is acknowledged that development cannot be left entirely to its own, but must be shaped for sustainability. Proposed strategies aim at a certain quality of the emergent result by shaping the interaction processes from which they emerge, while they concede that the particular result of self-organisation cannot be predicted or controlled. The question is: how can self-organisation be modulated as to bring about development that qualifies as sustainable?

– Jan-Peter Voß, Jens Newig, Britta Kastens, Jochen Monstadt, Benjamin Nölting: Steering for Sustainable Development, 2006 –

This chapter deals with the praxiological aspect of the new *weltanschauung*. It elaborates, in a first step, on the philosophical part of the praxiological aspects of POE that give a fresh perspective on human strategies. These philosophical guidelines serve, in a second step, as the foundation for the cross-disciplinary, system theoretical specification of

---

[a] I was nudged by Richard H. Thaler and Cass R. Sunstein's then new publication *Nudge – Improving Decisions about Health, Wealth, and Happiness* [2008], which made me take up this notion in a generalised manner. (And I owe thanks to Jack Curtin, an American friend, who nudged me by drawing my attention to the book.) I take the term *pars pro toto* for signifying the praxiological key concept of the new *weltanschauung*.

praxiology. This is known as Evolutionary Systems Design which, in turn, forms part of EST. Both on the level of philosophy (POE) and on the level of system theory (EST), principles are formulated that represent steps to a UTI.

3.1 A Fresh Perspective on Human Strategies

Column one of Table 2.3 is the next item of discussion.

The starting point is the "controllability vs. uncontrollability" dispute, i.e. the question whether or not humans are able to control events and entities in the world.

The answers depend on which way of thinking is applied. Two (activist) answers are in favour of controllability:

(1)   the practicist answer,
(2)   and the utopianist and the romanticist answer, which belong to the same category.
(3)   A third answer is in favour of uncontrollability. This represents the inactivist answer.
(4)   And there is a deliberately activist answer which provides the fresh perspective beyond the controllability vs. uncontrollability schism.

Table 3.1. Four strategies from the philosophical point of view.

	basic assumption	concepts used	means considered	feasible and desirable correlated
practicism	belief in blunt progress		expensive (brute force) intervention	every feasible is desirable
utopianism, romanticism	wishful thinking	"management"		every desirable is feasible
inactivism	passivism	"inviolability"	non-intervention	no match
deliberate activism	limited controllability	"participation", "partnership"	nudges	unity to-be-produced

The description of each strategy will include, after introductory remarks, the basic assumption on which the strategy is founded, the terms with which it argues, its attitude to interventions as a means of implementing the strategy, and the implied relationship between the feasible and the desirable (Table 3.1).

3.1.1 *Practicism*

A positive answer comes from modernism. Modernism is the ideology of modernity. Modernity is that age of human history in which a particular type of civilisational development is said to predominate. This mode of civilisation has its roots in the Christian-occidental mode of science and technology, whose innovations are seen as the driving force of society. Today, the western type of science and technology, the related industrial and computerised takeover of the natural world, and the resulting uniform culture of capitalism, democracy and human rights are the main features of modernity. This is referred to here as practicism (Table 2.2, line 1, and Table 3.1, line 1).

Belief in blunt progress. The conviction of modernism is that progress in science and technology is automatically translated into progress in society. This is demonstrably untrue and may be referred to as an illusion of omnipotence.

"Management". This is an optimistic view for those who are in power. It implies that everything can be managed, steered, planned: everything can be controlled totally, if there is the will to do so.

Expensive (brute force) interventions. Interventions aim to produce final, desired states. They functionalise cause-effect relationships such that the causes equal the initial states of departure and the effects equal the desired states of arrival. Interventions are linearly sequenced operations. Interventions may be expensive in that the means used are not optimally efficient, but they are effective in that they yield the desired result. It may require a major effort to put the means to work. And the means may also yield undesired results.

The feasible is desired. This modernist view may be traced back to the Bible. It can be termed "dominionism" because it aims at erecting a dominion over the world we live in. Accordingly, everything that can be made shall be allowed for. And, in principle, there is nothing that cannot be made (or at least only very few things). Nature, for example, can be changed to fit human whims. What is wishful, desired, is reduced to what can be made, what is feasible.

3.1.2 *Utopianism, Romanticism*

Another positive answer is given by the direct opposite of modernism, by fundamentalist anti-modernism, a term used here to describe the most fundamental crtiticism of Modernity. Notwithstanding, anti-modernism, in the form of fundamentalism, can be characterised by the same belief in intervening in the world and is thus another variety of the illusion of omnipotence. It shares with the former activism the "management" talk and allowance for expensive interventions. It differs from modernism perhaps only in emphasising the final cause because it prioritises values, ethics and morals that oppose Modernity.

Wishful thinking. Fundamentalist anti-modernism comes in two forms (Table 2.2, line 2, and Table 3.1, line 2).

(1) One is directed towards the future and seeks to implement a blueprint not substantiated by the actual space of possibilities. This is utopianism. As Karl Raimund Popper noted on several occasions, the attempt to bring heaven down to earth will most likely result in hell [e.g. 2005, 178]. The dictum "the revolution ate its children" is in the same vein.

(2) The other is directed towards the past and seeks to turn back the clock without, again, recognising the actual space of possibilities. That form of activism is reactionary romanticism because, in the past, there was supposedly a Golden Age when the world was in order.

The desired is feasible. Since utopianism and romanticism suppose that what they want to achieve can be made true, they project the desired onto the feasible (which contrasts with practicism).

3.1.3 *Inactivism*

A denial of controllability is put forward by a third train of thought (Table 2.2, line 3, and Table 3.1, line 3). It shares with utopianism and romanticism the critique of Modernity, but opposes any type of activism. It is anti-modernism in the form of the ideology of Postmodernity.

Passivism. It derives an illusion of impotence due to the experience of Modernity being confronted with many undesired results – side-effects in other domains of our world, local and far-distance effects, and short- and long-term effects – which are detrimental to our survival.

"Inviolability". The world is taboo. Nature, Creation, fellow humans are treated as inviolable, sometimes as holy.

Non-intervention. Passivism refrains from any intervention, from any action.

No match between the feasible and the desired. What would be desired cannot be brought into line with what can be made, because nothing can be made.

3.1.4 *Deliberate activism*

A hiatus, even a contradiction, between the standpoints of unlimited controllability (activism) and the standpoint of complete uncontrollability (inactivism) is typical of Modernity and its anti-modern and postmodern counter-movements. Both sides, however, are counterproductive. They do not ensure the unity of practice. They do not show a way to tackle the complex and global problems.

On the one hand, pursuing the path of modernity is not plausible (in the way that a simple increase in science and technology with the same economic drives and political framework conditions could qualitatively change the situation) if the current situation is due to a lower quantity of the same development. In this conservative variant, continuity becomes an absolute and the necessity for or possibility of a jump in quality is denied. Two views are possible: the global problems are either no problems at all or civilisation's development is in any case endowed with a problem-solving capacity that is sufficient to solve those problems. Neither case requires action to be taken.

On the other hand, the call for a U-turn would throw the baby out with the bathwater if it proposed something radically different here and now, without any consideration of past developments. It believes it would have to renounce modern science or technology, just as it would have to forego modern economy and politics. This radically utopian form of socio-political guidelines makes discontinuity absolute, and denies the possibility or necessity of the continuation of certain relationship structures in societal development; it dualises the bad reality and desired good to the point that every possible course of action becomes superfluous.

Apart from these two alternatives, there is a way out that stresses the possibility and necessity of both discontinuity and continuity in the scientific-technical development, which is enclosed in societal development. The global problems have their ultimate cause in socio-political developments, but are accelerated by scientific technological progress: any solutions must interconnect social and technological changes. Science and technology can only do justice to their original purpose – to alleviate suffering and generally make that life more pleasant – when they are no longer left to pursue their seemingly natural course. Instead of being left to their own dynamics, they should be deliberately put into operation after appropriate reflection and careful consideration. They should be managed with conscious control, i.e., when their programme is executed according to the ideals of humanity's survival in a future in which it is worth living. A constant control of the results of implementing the programme must also be instituted. Science must therefore carefully consider its technological consequences in society, must anticipate possible desired or undesired effects, and must conduct all appropriate readjustments or reorientations.

This kind of activism is not a practicism that guides action according to the maxim that all that is feasible shall be realised, thereby assuming that it is desired too. Nor is this kind of activism a utopian or romantic wishful thinking that holds that what is desired is feasible too. Both practicism and wishful thinking believe in total controllability and often result in expensive brute-force interventions. Nor is this kind of activism an inactivism that believes in total uncontrollability, condemns any kind of intervention and fails to reconcile the feasible and the wishful. Rather, it is deliberate activism that tries to ensure the unity of practice (Table 2.2, line 4, and Table 3.1, line 4).

Limited controllability. Deliberate activism therefore boils down to advocating a human strategy that is more modest, more decent, more cautious than those strategies characteristic of practicist, utopianist or romanticist activism. Deliberate activism, moreover, cannot compare with inactivism because the latter provides no strategy whatsoever.

What deliberate activism seeks to establish is a kind of permanent feedback mechanism that links the outcome of action to the reflection of its consequences, the goal being to inform the next required action. The current state must be repeatedly measured against the target state (and the target state need not to be fixed). This closely resembles what is intended by the institutionalisation of technology assessment, which provides an account of the impact of technology to inform technology design.

Precursors of this idea were formulated by Popper [1972, 2005], who developed a view of piecemeal engineering that, according to him, is a type of social engineering contrary to utopian engineering. The difference between piecemeal and utopianist action, disregarding the ideological ballast with which Popper loaded his idea [Avery 2000], lies in the method. As Popper concedes [2005, 158],

> The politician who adopts this [piecemeal] method may or may not have a blueprint of society before his mind, he may or may not hope that mankind will one day realize an ideal state, and achieve happiness and perfection on earth. But he will be aware that perfection, if at all attainable, is far distant and that every generation of men, and therefore also the living, have a claim...

By social engineering, Popper means something analogous to engineering, in its common-sense meaning [1972, 64-65]:

> Just as the main task of the physical engineer is to design machines and to remodel and service them, the task of the piecemeal social engineer is to design social institutions and to reconstruct and run those already in existence.

Another similar idea dates still further back to Robert K. Merton's essay *The Unanticipated Consequences of Purposive Social Action* [1936]. Today, some refer to it as the "law of unintended consequences", meaning that there are always consequences that are produced by action

– not by intention, not by foresight – and are often harmful. This idea, too, is a frustration of the hubris of control.

"Participation", "partnership". The phrase that adequately expresses the new relationship to the world is that of "taking part in a greater whole". It signifies self-interest should, as a particular entity, be in accord with the world's totality, i.e. be universal. Nonetheless, that whole need not to be understood as being holy.

Nudges. Instead of expensive interventions or non-intervention, nudges are the treatment of choice. Nudging is directly in between the poles of enforcement and reluctance to act. It sets cues that might or might not be taken up and trigger the intended outcome.

Bacon is sometimes mistakenly cited as being paradigmatic for torturing nature. Close scrutiny, however, reveals another attitude. Evidence is found in Bacon's *De sapientia veterum* – which was published in 1609 in Latin and comprises Bacon's re-narration of 31 myths of Antiquity together with his interpretations – that humans should seek to embrace Nature rather than fight her. Accordingly, Orphean wisdom by far outbalances violent pacification *á la* Hercules and makes stones and trees and animals gather around humans peacefully [1990].

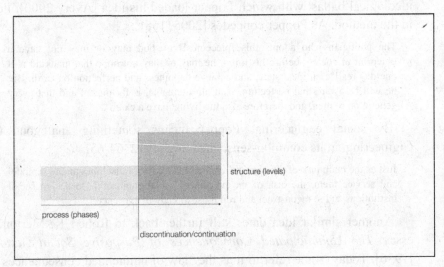

Figure 3.1. Multistage scheme.

Respect for Nature means, translated into today's language, obeying its laws. Changing the world is a permitted, and even recommended, if not unavoidable strategy, as long humans are aware of the restraints that are imposed by the functioning of the world.

Unity of the feasible and the desired. Deliberate activism takes responsibility for producing the unity of the feasible and the wishful. It does so by elaborating the ascendence from the potential to the actual and from the less good to the better.

This calls for considering not only the possible but also the desirable (which differentiates deliberate activism from practicism); conversely, we must also consider that which is not only desirable but also possible (which differentiates deliberate activism from utopianism and romanticism).

The approach is the critical aspect: it includes not only an account of the potential that is given with the actual, but also evaluates the potential which selects the desired. Thus, philosophy embraces a) an ascendence from the given potential to the actual that is to be established in the future and b) an ascendence from the "less good now" to the "better-then", which altogether yields the Not-Yet in the sense of critical theorist Ernst Bloch [1967].

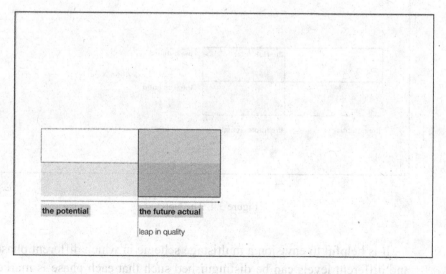

Figure 3.2. The ascendence from the potential to the future actual.

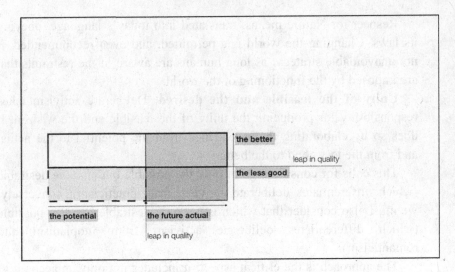

Figure 3.3. The ascendence from the less good to the better.

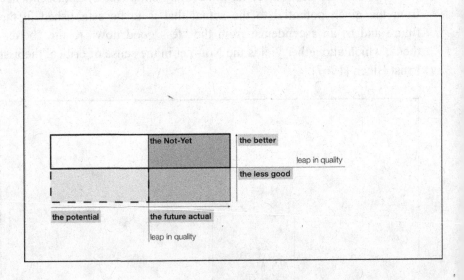

Figure 3.4. The Not-Yet.

It is helpful to envision a multistage scheme in which different phases and different levels can be distinguished such that each phase is marked by a leap in quality that introduces another level (Figure 3.1). Then, the potential is a space of possibilities rooted in the conditions of the actual in a certain phase on a certain level. The future actual is a realisation of a certain possibility from the space of possibilities, which adds a level to the existing ones (Figure 3.2).

However, this is not the whole story. A value has to be assigned to the potential to discriminate whether or not it is better than the actual *hic et nunc*. The potential can only be realised in the case of improvement. This is an ascendence from the less good to the better (Figure 3.3).

Altogether, a breakthrough to the Not-Yet is performed. The Not-Yet is then the future better (Figure 3.4). It is anchored in the present potential. Thus, it is said to flash up in the now. The flash-ups foreshadow the better future.

This type of activism may be defined as follows:

Deliberate activism. *Deliberate activism is that mode of human strategy that applies the Principle of Limited Controllability. The Principle of Limited Controllability is based upon the Principle of Unity-Through-Diversity.*

The **Principle of Limited Controllability** states: Human strategy is capable of controlling the consequences of action within a certain limit.

This limit is a constraint and an enablement in one – beyond the limit there is no possibility of control, while the limit gives room for control realisations that are sufficient for the survival of mankind and a good life for all.

Human strategy has to

(I)   check the consequences at each step;

(II)   accord the particular interest with the universal context;

(III) use nudges as a means of influencing events or entities;

(IV) reconcile – as Not-Yet – the feasible with the wishful.

The overall conclusions is that, *vis-à-vis* the world, actively positioned humans must deliberate the strategy required to change the world for the better. Deliberate activism is the praxiological instantiation of POE.

3.2 Dealing with Complexity

From a system theory perspective, dealing with complexity is located inbetween two poles – the pole of algorithmisation and the pole of nonprogrammability. This "algorithmisation vs. nonprogrammability" tension is the system theory perspective analogue to the above "controllability vs. uncontrollability" tension.

Algorithmisation means the process of developing clear-cut and unambiguous instructions that can be carried using computers as universal machines. To date, most technologies have been designed according to such a rule, but disasters increasingly render this view inappropriate. Nonprogrammability refers to this very experience that systems in nature and society do not obey that rule: they escape simulation and intervention and cannot be reproduced.

The above considerations define how the strategies categorised according to philosophical considerations appear from the angle of system theory. The praxiological divide between activism and inactivism transforms into a dirigism–*laissez-faire* divide in the system theoretical perspective (Table 3.2).

Table 3.2. Four strategies from the system theoretical point of view.

strategy in philosophical terms		strategy in cross-disciplinary terms		core idea
activism	practicism	dirigism	cybernetics, informatics	regulation
activism	utopianism, romanticism	dirigism	magical thinking	regulation
inactivism		laissez-faire		anarchy
deliberate activism		evolutionary systems design		decentralised context steering

"Dirigism" in this context refers to the conviction that strategies can be algorithmised and actions can be broken down into operations whose consecutive execution leads from the starting point to the planned endpoint. The algorithm provides the chain between the operations and ensures a linear *procedere* by prescribing which operation is to follow which operation. "*Laissez-faire*" describes the opposite conviction of nonprogrammibilty. And designing evolutionary systems provides the sublation of this tension.

Much like activism, which comprises two different strategies, dirigism comes in two varieties.

One is the transformation of practicism into a cross-disciplinary view, the other is the transformation of the twin of practicism, the projectivistic illusion of omnipotence.

3.2.1 Cybernetics, informatics

The first variety of dirigism can be exemplified by first-order cybernetics and informatics assumptions. The same features that hold true for practicism are valid for both first-order cybernetics and informatics assumptions. This includes the basic assumption, the concepts used, the means considered, and how the feasible is correlated with the desirable. First-order cybernetics and informatics are elements of engineering sciences and share the practical orientation of intervening in systems. Thus, the core idea involves regulation, which is as Klaus Krippendorff states [François 2004, 495]

> Any systematic (rule-like or determinate) behavior of one part of a system that tends to restrict the fluctuations in behavior of another part of that system.

The role model for a regulator is a thermostat.

> While both parts must lie in the same feedback loop, regulation involves this basic asymmetry: the regulator detects and responses to discrepancies from some expectation (criterion, goal) which is of an ordinality higher than the behavior so assessed and it computes the actions appropriate to keep the behavior to be regulated within desirable limits.

According to Norbert Wiener's *opus magnum* from 1948, cybernetics is the "field of control and communication in the animal and the machine" [François 2004, 145]. The British psychiatrist William Ross Ashby supported this idea by stating [François 2004, 145]

Cybernetics deals with all forms of behavior insofar as they are regular, determinate or reproducible...

Control is then a specific form of regulation that involves human intention. As Charles François annotates in his *International Encyclopedia of Systems and Cybernetics* [2004, 147],

control in its algorithmical sense reflects the more or less totalitarian mind and spirit (and possibly the scientific "ubris") of many mid-20th century leaders in most fields.

One exception is the famous cybernetician Stafford Beer. For him [François 2004, 146]

Cybernetics begins where the possibility of algorithmization of the controlled system ends.

Thus, it comes as no surprise that Luwig von Bertalanffy was quite skeptical of cybernetics. As David Pouvreau points out in his biography of Bertalanffy, the founder of the General System Theory defined the relation between the former and the latter in a lecture given at a conference in Toronto very soon after Wiener had coined the neologism "cybernetics" in the following way [2009, 124]:

Bertalanffy propounded that cybernetics is based on the concepts of feedback and homeostatic equilibrium. It develops the vision of essentially reactive systems, whose structural conditions are rigid and whose degree of organization can only increase when external "information" is provided: it is basically only a refinement of the behaviourist model of "stimulus and response", which it supplements with the concept of feedback. To this "mechanistic theory of systems" Bertalanffy opposed his general systemology as an "organismic theory of systems": its fundamental concepts are those of the "open system" (open to exchanges of matter and energy, and not only to "information") and of dynamic interaction; its emphasis is put on the states of non-equilibrium, more precisely on the steady state; and it develops the vision of primarily active systems, capable – due to their own dynamics – of organizing themselves progressively. Bertalanffy, perfectly in line with his "organismic" principles, conceived cybernetics as a particular field of general systemology: that which deals with mechanized systems whose characteristics of self-regulation rest only on fixed structural arrangements and which are thus "secondary" with regard to those "primary" ones, which result from the dynamic interactions between the components of the system.

Bertalanffy was so correct that his characterisation of the assumptions of cybernetics proved to be valid for second-order cybernetics as well (see next chapter).

Cybernetics already includes the study of information, as becomes clear from its emphasis on control and regulation. According to Beer [François 2004, 146]

> cybernetics studies the flow of information round a system, and the way in which this information is used by the system as a means of controlling itself...

Informatics then specialised to do this task even better. Informatics is the field of computer science and related disciplines encompassing artificial intelligence, cognitive science, social informatics and others. Its focus revolves around information processing in digital computers. Insofar as the technical aspect takes center stage, it is justified to refer to the assumptions underlying these disciplines as being technically reduced. In particular, systems in nature and social life are likened in their functioning to systems built by engineering science, so-called "information systems". The fallacy is to infer from artificial systems that real-world systems are to be handled in the same manner. This has been termed the computer metaphor of the brain (to begin with a small-sized problem) and as the computer metaphor of the universe (to end up with the largest imaginable problem).

3.2.2 *Magical thinking*

The second variety of dirigism is a type of magical thinking. It differs from cybernetics and informatics in that regulation is assumed to involve spiritual, supernatural pathways rather than material-energetic ones. "Magical thinking" goes beyond a psychological term describing certain states of the human mind, either in psychopathology or in developmental psychology. It is also an anthropological term signifying a state in the development of human thinking that precedes scientific rational thinking. Rain dance, voodoo and astrology belong to magical thinking. Cabalistics, as a play with numbers (see section 5.2.2), contributes to magical thinking as the following example illustrates. Peter Russell [1983], the first to speak of a global brain, emphasised that an increase in quantity is necessary for evolution to enable the emergence of a new quality. According to him, there is a magic number that, in the near future, by means of telecommunication, will reach a state of interconnectivity comparable to that of human brain cells. This number is 10^{10}, an order of magnitude characteristic for neurons building a human brain, as well as a number applying to the world population.

Magical thinking affirms algorithmic thinking, though the algorithm in that case may be esoteric and occult.

3.2.3 *Laissez-faire*

Inactivism, translated into the system theoretical view, yields what is known as *laissez-faire*. Here, too, what is true of inactivism is true of *laissez-faire*: passivism as a basic assumption, the use of the concept of "inviolability", decided non-interventionism with regard to means, and no congruency of the feasible and the desirable. *Laissez-faire* manifests an indifference, a lack of interest in changing, influencing, or modifying the systems that make up the world. The result: anomy or a kind of anarchy occurs in the place of dirigism's regulability idea. The assumption is irreproducibility rather than any visible or clandestine algorithmic pattern by which the systems can be governed. Events and entities cannot be programmed.

A well-known example for such a recently recognized, failed strategy is in the field of politics and its relation to economy: neo-liberal policy conjured the retreat of the state, the advent of liberalism, privatisation, and deregulation – unleashing the so-called free market forces. Actually, rather than retreating from the economy, the state provided and safeguarded the framework conditions that favoured profit-making. Certain pedagogical theories provide additional examples.

3.2.4 *Evolutionary Systems Design*

Just as the activism–inactivism divide can be overcome by claiming deliberate activism, the divide between dirigism and *laissez-faire* can be rendered obsolete by the claims of Evolutionary Systems Design.

Evolutionary Systems Design is the system theory instance of deliberate activism and its basic assumption of limited controllability, its "participation" and "partnership" concepts, its consideration of nudges and its imperative of producing the unity of the feasible and the desirable.

Evolutionary Systems Design ideas can be abstracted from the wide range of findings in systems research. This avoids becoming deadlocked at one of the poles of algorithmisability and nonprogrammability.

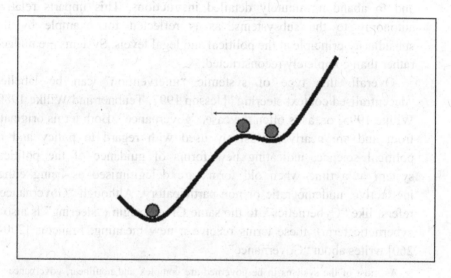

Figure 3.5. Landscape of system states.

These ideas encourage making use of the systems' dynamic. They stress the point that knowing about nonlinearity and sensitivity may help to choose those inputs that the system might take as cues for developments in the system's overall self-organisation process that are favourable to those who make the inputs. This may facilitate or dampen system processes. It is incorrect to assume that, if systems are self-organising – be it on the physical, biotic or social level of evolution – humans have no say and cannot influence them. The more humans know about the functioning of the dynamic of self-organisation in the specific case, the more they can act intelligently *vis-à-vis* that system and either side with or counter the system's self-organising capacity. This functions like jiujitsu. This is a smart strategy. Note, however, that systems are nudged rather than completely subdued, and humans can never be sure about the system's response in detail.

This situation can be visualised by drawing a landscape of system states in which equilibria are marked as those points lying at the bottom of valleys. Accordingly, plans can be plausibly devised that, with little effort, can nudge the system to assume another state (Figure 3.5).

It is also important to influence only the general set-up of the system and to abandon minutely detailed instructions. This imparts relative autonomy to the subsystems, as is reflected for example by the subsidiarity principle at the political and legal levels. Systems are nudged rather than completely reconstructed.

Overall, this type of systemic "intervention" can be labelled "decentralised context steering" [Jessop 1997, Teubner and Willke 1980, Willke 1995] or, as is often the case, "governance". Both terms originate from and are nearly exclusively used with regard to policy and in political science, indicating new forms of guidance of the political system at a time when old forms are delegitimised as being either ineffective. undemocratic or non-participatory. Although "Governance" refers, like "Cybernetics", to the same Greek origin ("steering" is also a cybernetic term), these terms receive a new meaning. François [2004, 260] writes about "Governance":

> As most of the systems to be governed are complex and nonlinear, governance should not be taken as a simple (nor even complicated) deterministic procedure.

The following definition of designing evolutionary systems is proposed here:

Evolutionary Systems Design.[b] *Evolutionary Systems Design is that mode of strategy that applies the Principle of Decentralised Context Steering (Governance). The Principle of Decentralised Context Steering (Governance) is based upon the Principle of Limited Controllability as well as upon the Principle of Self-Organisation.*

---

[b] For another definition by Ken Bausch, see [François 2004, 218].

The **Principle of Decentralised Context Steering (Governance)** states: Human strategy is to deal with self-organising systems and has to acknowledge irreproducibility. Nonetheless, that strategy is capable of influencing (I) the process of self-organisation and (II) the self-organised structure of the system in question.

In particular, the Principle of Decentralised Context Steering (Governance) consists of two principles:

(I) The **Principle of Making Use of the Dynamics** states with regard to diachrony: Choose inputs to facilitate or dampen the dynamics of self-organisation!

(II) The **Principle of Making Use of the Architecture** states with regard to synchrony: Grant relative autonomy and shape the general set-up of the self-organised architecture only!

Summa summarum, nudges are the key element of Evolutionary Systems Design into which the new philosophy's praxiology translates. Evolutionary Systems Design is smart, fuzzy, indirect control by irritation of the system to be governed rather than totally programming it or completely giving up regulation. Ervin Laszlo points out, regarding the relationship between humans and nature [2008, 15],

> that not only is nature a dynamic system capable of rapid transformation but humanity is also. When such a system nears the point where the existing structures and feedbacks can no longer maintain its integrity, it becomes ultrasensitive and responds even to the smallest provocation for change.

Thus, "design" means shaping systems rather than inventing them, constructing them from the scratch, or even deconstructing them.

Chapter 4

Chaosmic Metasystem Transitions and Suprasystem Hierarchies

> The old universe was a perfectly regulated watch. The new universe is an uncertain cloud.
>
> [...] uncertainty which is inevitably ours, who are peripheral observers, limited in our senses, deformed in our intellect, ignorant of most what goes on in space and of all that will unfold in time, may also be, to boot, the uncertainty of the universe itself, which does not yet know what is going to happen to it [...]
>
> – Edgar Morin: Method 1, 1992 (Nature de la nature, 1977) –

Having dealt with the praxiology of the new paradigm, the next step is to examine its ontology. This chapter is also divided into two sections, the first one, again, deals with the POE issue of ontology, the second one deals with ontology from an EST point of view. The former provides a fresh perspective on the real world, in particular on how to model the interconnectedness and interdependence of events and entities in the world, and the latter specifies those insights in determinacy and indeterminacy for the dynamics and architecture of evolutionary systems, which is referred to as Evolutionary Systems Ontology. Both sections, again, formulate principles for modelling – yielding two further steps toward a UTI.

4.1 A Fresh Perspective on the Real World Image

This section elaborates on column 2 in Table 2.3. The responses to the question of "determinacy vs. indeterminacy" rest upon ways of thinking and comprise four different options. Two of them favour strict determinism, one favours indeterminism, and one favours the sublation of this contradiction by proposing a determinism that is less than strict. These are:

(1) preformationism and atomism,
(2) teleologism and holism,
(3) indeterminism,
(4) less-than-strict determinism.

Each option is described in terms of its basic assumption, the metaphors it relies on, the causes taken into consideration, and the correlation of the one and the many, that is, old and new as well as of parts and the whole (Table 4.1).

Table 4.1. Four conceptions of the interconnectedness of the universe from the philosophical point of view.

	basic assumption	metaphors used	causes considered	the one and the many (old/new, parts/whole) correlated
preformationism, atomism				nothing new, sum of parts
teleologism, holism	cosmos	"clockwork"	necessity	drain, independent whole
indeterminism	chaos	"clouds, bolts from the blue"	no necessity	everything may happen
less-than-strict determinism	chaosmos, propensities, habits	"great oaks from little acorns"	necessity and contingency	unity to-be-produced

4.1.1 *Preformationism, atomism*

Mechanicism, or better, mechanicist determinism, is the ideal toward which mainstream thinking in (natural) sciences tends. This is materialism in that it denies ideal causes. All phenomena are explained by reducing effects to material causes that are deemed sufficient to produce those effects.

Determinism is the view of determinacy and indeterminacy in real-world causal relationships. It is about whether causes determine strictly or not at all or something in between.

Note that the notion of "determinism" as used here might differ from the commonly used term. Traditional science tends not to distinguish between "determinism" and the "principle of causality" and thus to conflate both. The principle of causality holds that there is no event that is not caused, i.e., every event is an effect of a cause. The underlying assumption is a closed chain of causes and effects. "Causality", however, signifies that interaction between events which is direct. "Determinism", in contrast, is about interaction between events, be it direct or "indirect". So-called "indirect" interaction refers to laws governing that part of the interaction that is universal and necessary, and to chance, which governs that part of the interaction that is not universal but particular and not necessary but random. Removing chance from the principle of causality yields indeterminism. The interpretation of causality and determinism given above involves random events that are nevertheless caused [Hörz 1962, Hörz 1971, 208, Fuchs-Kittowski 1976, 178-187].

The mechanistic view is associated with the names of Isaac Newton and Pierre-Simon Laplace.

Cosmos. Newton's mechanical perception of the world was based on three principles [Gerthsen et al. 1995, 13, Fleissner et al. 1997]:

(1) The principle of inertia: a body on which no forces are exerted moves constantly in a straight line.

(2) The principle of action: If a force F is exerted on a body of mass m and velocity v, the impulse of the body, mv, is changed, such that $d/dt\,(mv) = F$.

(3) The principle of reaction: If the force F which is acting on a body has its origin in another body, exactly the opposite force $-F$ is acting on the latter.

Newton's classical mechanics used the concept of causality in an elementary way. If a force acts on a body, by the principle of action the velocity of the body is changed in a unique way. The body is accelerated proportionately to the force exerted.

These principles imply the unique determination of the effect based on a known cause. Newton's writings became the prototype for scientific reasoning, establishing determinism by eradicating types of cause other than efficient cause.

The basic assumption is thus a universe that is completely ordered, that is, a cosmos in the original sense of the word: every event and every entity is strictly determined as a determinate effect of a determinate cause in a bijective mapping. If cause and effect are related in such a way that each cause is related to one, and only one, effect, then determinism is held to be complete [Heylighen 1990] (Figure 4.1).

In this sense, *causa aequat effectum*, or – as Newton's dictum was interpreted elsewhere [Fleissner et al. 1997] – *actio est reactio*. The mathematical function provides a tool that seems to be guarantee calculable results.

"Clockwork". This mechanistic worldview was made explicit by the well-known idea of Laplace that a demon who knew the world formula plus all data describing a certain state of the universe would be capable of predicting and retrodicting any state of the universe. In Popper's terms, this may be referred to as the clockwork view of the universe [Popper 1966].

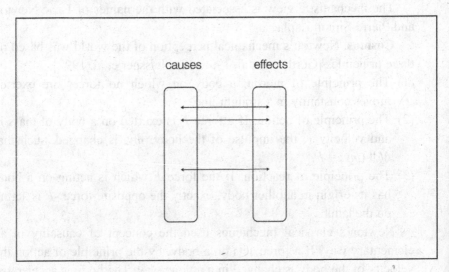

Figure 4.1. Cause-effect mapping in strict determinism.

Necessity. The cause-effect-relationships considered in that context are all of the same nature. All convey necessity, all are necessary and do not distinguish between necessity and contingency, between necessary and contingent causations.

Nothing new. Mechanical determinism comes in two varieties depending on whether the focus is on process, on time, on development, or on structure, on spaces, on states. The first aspect is termed "diachronous", the second "synchronous".

In the diachronous aspect, mechanical determinism assumes evolution – in accord with the etymology of its word – to be an "e-volution", i.e., an unwrapping of something that was already there before it is unwrapped, and not so much the appearance of something which has not been there before. Hence there cannot be anything new, because everything can be reduced to something old: there is no new thing under the sun. This variety of mechanicism is preformationism.

Sum of parts. In the synchronous aspect, mechanical determinism reduces wholes to their parts, which is known as variety of atomism – albeit not in the physical sense. The whole is said to be completely determined by its parts. There is no whole that is "more than the sum" of its parts. The world is explained by summarising all its parts.

4.1.2 *Teleologism, holism*

The opposite of the mechanistic view is idealistic mysticism. This determinism may be as strict as that of mechanicism; the difference is that the causes do have an idealistic element. Some of the humanities tend to be biased in this manner.

Regarding diachrony and synchrony, two varieties can be distinguished again.

Drain. The evolution of something that is said to evolve seems to be strictly governed by a *telos* that determines current developments out of the future. It is a pull-model, in contrast to the push-model of mechanicism. This is known as teleology.

Independent whole. Moreover, wholes seem to exert such a strong pressure on their parts that those parts cannot influence the wholes. The wholes exist, independently. This is called holism.

4.1.3 *Indeterminism*

The opposite of the classical view of both mechanistic materialism and idealistic mysticism is dualistic mysticism.

Chaos. It denies that effects are caused, and this holds that there is no sense in ascribing cause-effect-roles to events or entities. From this perspective the world is heterogeneous, fragmented and disintegrated, and it falls apart in disjunctive sets. It is chaos in the original sense of the word: disorder.

"Clouds", "bolts from the blue". Popper likened this view, in contrast to the classical one for which the metaphor is "clockwork", to that of clouds. Becoming and being is compared with clouds: events and entities are like bolts from the blue.

No necessity. Interactions contain no features of necessity. What happens or exists does so by accident.

Everything may happen. It is no surprise that every event and every entity is a surprise. Dualism overlooks continua and neglects both the old and parts, thus avoiding considerations that dichotomise old and new or parts and wholes. Old and new do not depend on each other; neither do parts and wholes. Evolution is as undetermined and history as arbitrary as the order and the logic of the structure.

4.1.4 *Less-than-strict determinism*[a]

The classical view of determinacy in mechanistic materialism or idealistic mysticism and the nonclassical view of indeterminacy in dualistic mysticism create a tension between extremes, none of which is promising. Only a postnonclassical view of determinacy that goes hand in hand with indeterminacy may help.

Chaosmos, propensities, habits. Unity of reality can be envisaged by recognising that deterministic events and entities are merely a special case of events in the universe. This is because deterministic events and entities occur only with objects that are not agents. In the case of

---

[a] I used this wording thanks to Hans Peter Buetow, a native speaker who corrected my English style, first in [Hofkirchner 2001].

subjects, that is, subjective agents, events and entities are not strictly determined: the effect is not predictable because it is subjective agency that intervenes in the chain of cause and effect and introduces a degree of freedom that cannot be forced into a single alternative.

According to the POE point of view, the difference between subject and object is that a subject is capable of determining itself whereas an object is not. An object is something that is determined by something that is not itself. Being a subject supersedes being merely an object. While an object has no possibility of showing agency in ways different from merely reacting to external determinants, a subject is capable of responding in its own, unequivocal way. A subject can make use of degrees of freedom, of freedom of choice, of choice between options – all of which it has at its disposal and thus makes the internal determine. It may object to external determinants in a subjective way that objects are incapable of. Thus, there is objective indeterminacy as to subjective determinations.

Not only humans display this kind of subjectiveness. Making something subject to oneself (which makes oneself a subject) undergoes a process of unfolding. This allows us distinguish between different types of subjects in the world we inhabit according to the degree of subjectiveness they manifest. The minimal unit of subjectiveness is something that is provided with a minimal quantum of degrees of freedom to display agency. This something is the most rudimentary and most primitive subject. It differs fundamentally from being an object, that is, something that does not have options for reactions. It is already a kind of somebody – albeit not a human one, and also not a living one.

Subject-object-dialectic paints a new picture of the world: it is neither cosmos nor chaos, but bears features of both; it is "chaosmos" – a term coined by Morin [1992, 53-64]. When Popper wrote about propensities in one of his last publications, he had in mind the very same idea of a universe that is inclined to exhibit some properties rather than being governed by strict natural laws [1990]. The American pragmatist Charles Sanders Peirce stated that evolution takes habits [2000]. All of this is meaning: we live in a world in that indeterminacy inheres besides determinacy, that is open to the future, and that leaves room for subjects to decide.

There is no mechanistic transformation which turns the cause into the effect. There remains a gap in quality between cause and effect which is bridged in a mechanical way. Thus, *causa non aequat effectum, actio non est reactio*. One cause might have one effect or another, and different causes might have the same effect as long as all these relations are possible (Figure 4.2).

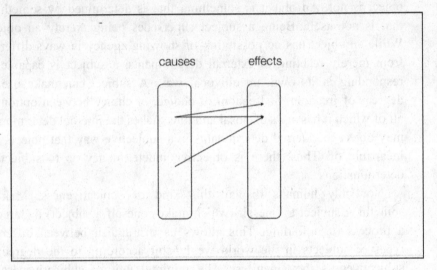

Figure 4.2. Cause-effect mapping in less-than-strict determinism.

"Great oaks from little acorns". Since subjects of all kinds interfere in causal relationships, the effect needs not to liken the cause. It can be different in size; it can be big but also small. At any rate, it is different in quality.

Emergentist philosophy, as developed for instance by C. Lewis Morgan and summed up by David Blitz [1992] in a book on *Emergent Evolution*, holds that effects which do not "result" from causes, that is, which are not "resultant" but "emergent", cannot be "reduced" to their causes. In this case, causation is only a necessary constraint, but not a sufficient one, as would be the case in mechanistic causation.

Hence, based on the concept of emergence, we have on the one hand the opportunity to stick to the principle of causality. This means that there is nothing which was created out of nothing (let's neglect the origin of the universe here). On the other hand, enough openness remains to let novelties arise which did not exist before.

Necessity and contingency. There is no determinacy without indeterminacy and no indeterminacy without determinacy. This calls for only a few necessary causal relations, whereas most of them are contingent, i.e. not strictly determined in detail.

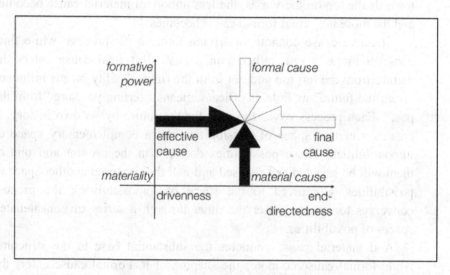

Figure 4.3. Aristotle's four causes revisited.

Aristotle recognised four types of causes: the effective (*causa efficiens*), the final (*causa finalis*), the material (*causa materialis*) and the formal (*causa formalis*) one. In a striving for scientific standards that avoided resorting to the supernatural, post-medieval science abandoned the latter three causes. Nonetheless, it is worth reconsidering all four types of causes without the need to resort to the supernatural. We can sort them into two pairs of opposites and arrange them on two continuum scales that stand orthogonally to each other (Figure 4.3). One axis shows the processual, diachronous dimension of events and goes from drivenness to end-directedness, another shows the structural, synchronous dimension of entities and goes from materiality to formative power [Brunner et al. 2003]. We can arrange the effective and final cause on the first axis and the material and formal cause on the second one in the following way: effective cause enters the picture from the left and final cause, as opposed to effective cause, is directed to the right. This means: the further we move to the right on the x-axis, the less important effective cause becomes and the more important the final cause; material cause enters the picture from the bottom and formal cause, as opposed to material cause, is directed to the bottom. This means: the more we move towards the top on the y-axis, the less important material cause becomes and the more important formal cause becomes.

Effective cause connotes a driving force in the process, while final cause connotes a pull rather than a push. But final cause enters the picture from the left too and not from the right. Finality means influence "from the future" as little as efficacy means exerting pressure "from the past". Each process paves the way for the future by its own history. It creates a certain space of possibilities and a complementary space of impossibilities. Those possibilities do exist in the present and one of them will be selected and realised and will then open up another space of possibilities. Compared to the space of impossibilities, the process converges to one end after the other through a series of concatenated spaces of possibilities.

And material cause connotes the substantial base in the structure, while formal cause connotes the shaping of it. Formal cause enters the picture from the bottom too, though its direction is top-down. It does not fall from heaven. Formality means influence "by mind" as little as materiality means exerting pressure "by matter". Each structure bears the stamp of how its constituents compose it. The constituents produce what they constitute by producing constraints as well as enablers which represent the form.

The further we move in the diagram from point zero to the right or to the top, the more contingency ensues and the less necessity is significant.

Having made these assumptions, we can identify subjects of all kinds and locate them according to the axes (Figure 4.4).

(1) The primordial stage of subjectiveness is given if ends are realised and forms are assumed (that help realise the ends) – these events or entities might be called "proto-subjects"[b].

---

[b] Francisco Salto helped me to name the graduation of different types of subjectiveness.

(2) Another stage may appear if ends that are realised are implicit, that is, built-in in these kinds of subjects and these kinds of subjects are able to perpetuate the forms they assume – these events or entities might be called "quasi-subjects".

(3) A third stage may be reached if ends are made explicit, that is, if they are decided by the subject and if the subject is able to create forms (which it perpetuates). Here, then, the term "subject" might apply in the full sense of the word as far as our knowledge allows.

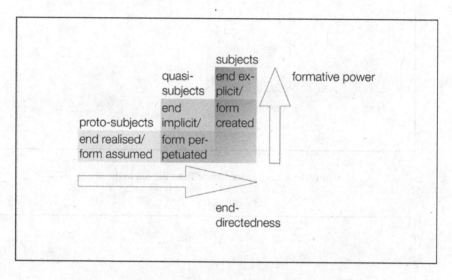

Figure 4.4. Stages of subjectiveness.

Unity of the one and the many (old and new, parts and whole). Less-than-strict-determinism is not a preformationism according to which evolution is merely an unfolding of something that already exists, and is not an atomism according to which wholes can be reduced to their parts. Both preformationism and atomism reduce contingency to necessity. Less-than-strict-determinism is not teleologism or holism, both of which project contingency in the form of a not existing goal or a contingent whole onto necessity. Finally, it is not a dichotomism of necessity and contingency. Rather, it attempts to work out the unity of necessity and contingency by a unity of the one and the many.

The old is something actual plus the space of possibilities rooted in the actual (Figure 4.5; non-filled areas mark the space of possibilities).

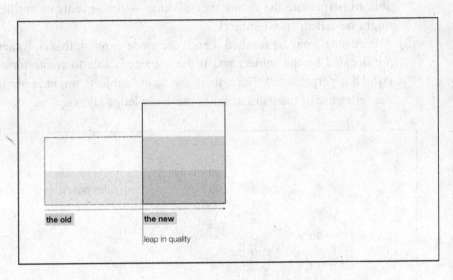

Figure 4.5. The ascendence from the old to the new.

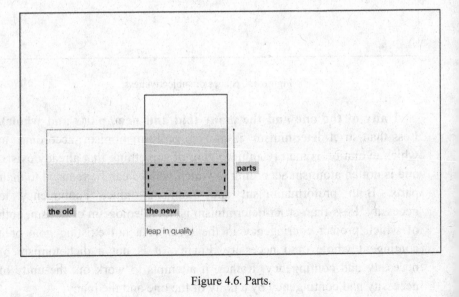

Figure 4.6. Parts.

If the old is replaced by something new, this new is the actualisation of one of the possibilities given with the old. This makes the new another actual which opens up another space of possibilities.

This actualisation shows both determinacy and indeterminacy. First, it is determined insofar as it cannot come true unless it is a possibility that belongs to the space of possibilities provided by the old. This is how necessity works. Second, it need not to be determined which of the possibilities comes true. This is how contingency comes into play.

The old provides for continuity in that the parts that produce the new are reworked from the old. The parts connect the new with the old (Figure 4.6).

In turn, the new whole produces its parts. It is the new whole that reworks the old (Figure 4.7).

The parts–whole relationship too combines determinacy and indeterminacy, necessity and contingency. In neither direction does the cause strictly determine the effect – not from the parts to the whole nor from the whole to the parts. This is because parts and the whole each possess subjectiveness and degrees of freedom.

Those parts belonging to a specific whole reflect this fact by possessing (at least) one property which they do not possess when being not part of this whole. At the same time, they are not completely absorbed in sharing that property. They have (at least) one other property which also makes them distinct. Thus, real-world parts are neither pieces or fragments that can do without the whole (just by taking away their property as part) nor are they instances of the whole (which means they share all properties of the whole). In turn, the whole possesses at least one property that it does not share with any of the parts.

Altogether, the ascendence from the old to the new and the ascendence from the parts to the whole accomplish the ascendence to a complex that unites the one and the many (Figure 4.8). The complex is a new variety that diverges from all that existed before. It branches off when adding another refinement to the old. By doing so it encapsulates varieties that lead back from the branch to the old. The complex integrates with the one by integrating the many as parts.

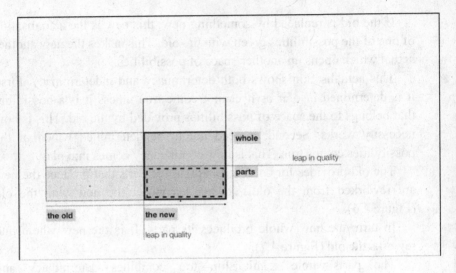

Figure 4.7. The ascendence from the parts to the whole.

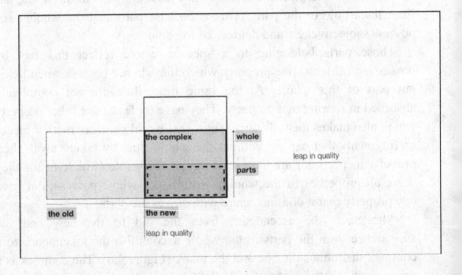

Figure 4.8. The complex.

This calls for defining less-than-strict determinism:

Less-than-strict determinism. *Less-than-strict determinism is that view of the interconnectedness of real-world events or entities that applies the Principle of Less-than-Strict Determinacy. The Principle of Less-than-Strict Determinacy is based upon the Principle of Limited Controllability.*

The **Principle of Less-than-Strict Determinacy** states: Determinacy includes indeterminacy.

Determination contains degrees of freedom.

The world

(I)   is open to the future and perfused with subjects of all kinds;

(II)   gives rise to emergent phenomena;

(III) is connected through an interplay of effective, final, material, and formal causes;

(IV) represents a unity of the one and the many, that is, the complex.

To recapitulate, less-than-strict determinism does not argue for no determinacy at all or that the clockwork view has to be replaced with a clouds view. It does not mean that anything goes. It only admits that nature itself is capable of spontaneously producing events and entities that are not describable in a mechanistic way. Besides and beyond clear-cut, one-to-one cause-effect-relations, there are also more flexible causal connections in the real world. In fact, the latter be more important and greater in number.

Accordingly, the thesis of less-than-strict determinism not only opposes the thesis of strict determinism but also leads to a new understanding of determinism which includes strictness as correct under certain conditions only.

4.2 At Home in Complexity

The philosophical part of the new ontology has to be transferred to the system theoretical perspective. Here the determinacy–indeterminacy divide can be translated into a mechanicity–spontaneity divide. While "mechanicity" means the conviction that systems are by nature mechanical (algorithmic programmes are applicable), spontaneity refers to the irreversibility of systemic developments and to the irreducibility of systemic build-ups. Accordingly, mechanicism and spontaneism form the two opposing conceptions of the interconnectedness of the universe from the system theory angle (Table 4.2).

Table 4.2. Four conceptions of the interconnectedness of the universe from the system theoretical point of view.

conception of interconnectedness in philosophical terms		conception of interconnectedness in cross-disciplinary terms		core idea
deter-minism	preformatio-nism, atomism	mechani-cism	evolutionism, modularism	
	teleologism, holism		creationism, structuralism/ functionalism	mechanical relations
indeterminism		spontaneism		anomy
less-than-strict determinism		evolutionary systems ontology		metasystem transition, suprasystem hierarchy

As with determinism, mechanicism varies between reductionistic and projectivistic forms.

Once again, the integrativist way of thinking helps overcome the cleft of the mechanistic and the spontaneistic views by depicting an evolutionary systems ontology based upon less-than-strict determinism.

4.2.1 *Evolutionism, modularism*

Mechanical systems work along the line of strict determinism. Péter Érdi writes [2008, 4]:

> Mechanistic reductionism suggested that the universe, including life, were considered as "mechanisms". Consequently, understanding any system required the application of the mental strategy of engineering: the whole system should be reduced to its parts. Knowing the parts was thought to imply the complete understanding of the whole.

The thesis of strict determinism (purporting mechanical, that is, simple, and not complex, systems) can be characterised as follows [Heylighen 1990, Weingartner 1996, 187-189]:

(1) Given a system, inputs and outputs are related such that each input is related to one, and only one, output. The system transforms the input into the output by way of a mechanism which can be conceived of as a bijection. If you call the input "cause", and the output "effect", you may state that equal causes have equal effects and distinct causes have distinct effects.

(2) Small changes in the causes lead to small changes in the effects.

(3) There are only repetitions. Each state of a system will return in the future; more precisely, known as Henri Poincaré's recurrence theorem, each state of a mechanical system will return if not to its initial state, at least arbitrarily close to it.

4.2.1.1 *Evolutionism*

Concerning diachrony, "evolutionism" depicts the mechanistic account of evolution, addressing the origin of species or any increase of complexity in a subsystem of the universe. This is the way this term is also used today by representatives of intelligent design, by which the scientific branch of creationism tags their opponents on the Darwinism side.

Not only presuppositions of necessity–chance mechanics in Darwinism belong to evolutionism, but also the contention that self-organising systems function as nontrivial automata and theses underlying the so-called deterministic chaos and cellular automata.

Darwinism

Darwinism – understood here as the mechanistic interpretation of Darwinian theory – is a good example for how evolution is modelled according to the functioning of mechanical relations. According to Darwinism, it is not the living system that would be an agent in evolution, let alone the environment that would be an agent that carries out the selection. However, by foregoing agents, no increase in complexity can be modelled. The classical example is the monkey-typewriter argument, which states that the time required for a monkey typing on a typewriter to end up with a text of William Shakespeare by far exceeds the time it took our universe to evolve so far.

The so-called molecular-biological dogma of Neo-Darwinism states that there is only one direction of causation from the genotype to the phenotype and not *vice versa*, more precisely, that causation follows the line

DNA $\rightarrow$ RNA $\rightarrow$ protein.

This has become softened in recent years not only due to the identification of an enzyme called reverse transcriptase by which RNA can be copied to DNA, but also due to findings in epigenesis that ascribe a vital role to factors in the environment of the developing organism [Ellersdorfer 1998]. Thus, genetic determinism has lost its attraction. Random mutations apparently do not occur at a constant rate. Rather, in times of increased stress, the probability of changes in the genome can rise by a factor of 10^4 [Bauer 2008, 95]. This is a plausible explanation for the fact that nosocomial germs did not lose their fight against antibiotics and go extinct. Transposable elements have been discovered that are responsible for restructuring the genome.

Nontrivial systems

Von Foerster is known for his distinction between trivial and nontrivial systems. A trivial system transforms an input x into an output y via an invariable relationship f, so that "a y once observed for a given x will be the same for the same given x later." The function f is said to be determined analytically, that is, an observer "simply has to record for each given x the corresponding y" [Foerster 1984, 9-10]. A nontrivial system differs from a trivial one in that

> a response $\{y\}$ once observed for a given stimulus $\{x\}$ may not be the same for the same stimulus later,

because it has at least one internal state z

> whose values co-determine its input-output relation (x,y). Moreover, the relationship between the present and subsequent internal states $\{z_t, z_{t+\Delta t}\}$ is co-determined by the input (x).

According to von Foerster, a nontrivial system is nontrivial only because the observer is faced with a nontrivial problem when trying to determine how the system works. Ontologically, however, there is no difference between trivial and nontrivial systems. Both types of systems can behave strictly deterministically. Once the mechanism of the function f_y and f_z of a nontrivial system is fixed, its output y, given an input x, is unambiguously determined [Hügin 1996, 128]. Thus, von Foerster's hidden ontology turns out to be still mechanistic.

Deterministic chaos, cellular automata

Another instantiation of mechanical systems is that which relies upon deterministic chaos or cellular automata. These systems are according to Heylighen [1989, 378]

> microscopically deterministic systems that behave in a completely unpredictable way when considered from a macroscopic viewpoint [...] whose local dynamical rules are completely deterministic but for which there is no global algorithm allowing to predict their overall evolution without computing all the individual, microscopic transitions from the given initial state.

They are said to exist not only in computer simulations but also in the real world. The solar system – apart from being an n-body problem that has no analytic solution because it consists of more than two bodies that interact with each other according to gravitation forces – was mentioned as a candidate for a deterministic chaotic system because, also here, the sensitivity to initial conditions is given. Epilepsy and heart beat rhythm are other candidates. Érdi states [2008, 87], chaos

> is a nonperiodic temporal behavior generated by purely deterministic mechanisms (in reality) and algorithms (in models).

Also, cellular automaton patterns are found in nature, e.g., in forming pigment patterns, swarms and other phenomena.

4.2.1.2 *Modularism*

Concerning synchrony, "modularism" (in social sciences: individualism) signifies the assumption that the behaviour, state or structure of a system can be reduced to the elements, their properties, or their interaction.

John W. Holland, the father of genetic algorithms, proposed to model emergence in hierarchical systems by using the modules which compose the system. However, the linking of the modules follows strictly deterministic rules [1998].

(Methodological) individualism

In social science, so-called methodological individualism presupposes the reduction of macro-level phenomena such as social relationships that characterise societies to micro-level phenomena due to individuals. Economic rational-choice theories represent one prototype.

4.2.2 *Creationism, structuralism/functionalism*

Mechanical systems may also be an inspiration when infusing teleological or holistic philosophical assumptions into modern systems thinking.

4.2.2.1 *Creationism*

As regards evolution, "creationism", like "evolutionism", is meant as a generic term; it does not only denote the ideology of the specific anti-scientific religious movement that fights against generalisations of evolutionary findings. "Creationism" thus comprises all forms of systems thinking with a teleological touch.

Examples range from the Gaia hypotheses to "intelligent design" to a variety of computer models using optimisation strategies such as backward propagation in neural networks, genetic algorithms, and fitness landscapes.

Optimising Gaia

The ideas of the engineer James Lovelock concerning the cybernetic feedback loops on planet Earth that preserve it as habitat for life have been interpreted in a New Age way. Accordingly, the mythological name "Gaia" that Lovelock attached to Earth's mechanical circles is taken seriously and not as a metaphor. In this interpretation, Earth itself is a living system and assumes teleological features. But also a less extreme formulation of the Gaia hypothesis includes an end toward which evolution is presumably heading. In the words of Peter Ward, who wrote *The Medea Hypothesis* [2009, xviii-xix], this formulation

> says that not only has and does life maintain 'habitability' for itself (albeit unconsciously, simply as an inherent property of itself), but it actually improves conditions by changing such factors as planetary atmospheric and oceanic chemistry, the cycling of elements through the biosphere, and the availability of nutrients to levels *more favorable for life*.

Intelligent design

An even more presumptuous theory is that of intelligent design which is a remake of a traditional proof for God's existence and takes advantage of the shortcomings of mechanistic Darwinism, thereby camouflaging that its own assumptions are at least as mechanistic as those of its opponent.

According to von Foerster, humans are prone to construct machines by trivialising systems, that is, by making them trivial systems whose behaviour responds to human expectations because they are built along deterministic rules [François 2004, 638]. In the very same way, an intelligent designer had to build her systems such that they work according to the designer's expectations. This design process then might be seen as revealing teleological features.

Backward propagation, genetic algorithms, fitness landscapes

In his book *The Blind Watchmaker* [1986], Richard Dawkins demonstrated that the Darwinist mutation–selection principle is sufficient to account for changes when modelled by a computer programme that simulates random mutations but includes evaluations of how much the mutation matches with the preset goal and selects the one with the best match for the next mutation step.

In the same year, Bernd-Olaf Küppers published the same argument in his book *Der Ursprung der biologischen Information* [1986]. These programmes work as optimisation strategies, which means a goal has to be imputed to them.

In the case of programmes, the goals are imputed by those who run the programmes. Who, however, would impute goals in the case of real-world systems?

Numerous programmes function with the same principle: backward propagation in neural networks, genetic algorithms invented by Holland [1975], evolutionary programming with fitness landscapes, and others.

4.2.2.2 *Structuralism, functionalism*

In social sciences and humanities, the holistic preference for mechanical systems leads to the influential streams of structuralist and/or functionalist (system) theories. In all of them, structures or functions that designate the macro-level are prior to the acts of actors on the micro-level. They have been criticised for being too strictly deterministic.

4.2.3 *Spontaneism*

Mechanical relations typical of mechanical systems and the ensuing world view are denied pursuant to the great narrative of Postmodernity, which posed that there is no great narrative. In line with the nonclassical uncontrollability and anarchistic *laissez-faire* concepts, indeterminism is expressed by the postulate of the existence of spontaneous forces in humans, life or nature as a whole. The result is a kind of "anomy" of real-world systems because they are not subject to natural laws.

Idealistic interpretations of emergentism are spontaneistic in that they absolutise the difference between the current development and the past development (thus denying path dependency) or between different systemic layers (thus extinguishing hierarchical propagation).

4.2.4 *Evolutionary Systems Ontology*

As science has unravelled the processes and structures of the natural world, mechanical relations and strict determinism, which are prevalent in the clockwork view of the universe, hold solely for systems at or near thermodynamic/chemical equilibrium. They do not, however, hold for systems exposed to fields in which the uneven distribution of energy density exceeds a critical level. Such field potentials force energy to flow in non-linear and interdependent ways. Under such conditions, the systems show self-organisation, that is the build-up of order out of fluctuations via dissipation of entropy.

The ontological EST considerations that follow in this subsection are presented in the same order as the issues had in subsection 4.1.4 about less-than-strict determinism. Hence, the issues here deal, sequentially, with the translation of

(1) the basic assumption of propensities and habits of chaosmos into the concept of complex systems,

(2) the metaphors about emergence into the language of system dynamics,

(3) the concept of contingent causation after Aristotle with the rise of subjectiveness into the view of system trajectories along with system isomorphies and system agency,

(4) the correlation of both old and new and parts and whole into the EST core idea of a stage model – path-dependent nestedness with metasystem transitions and suprasystem[c] hierarchies, the latter seen from an intrasystemic as well as an intersystemic angle.

4.2.4.1 *Chaosmic complexity*

Self-organising systems are complex systems. Among the large number of diverging definitions of complexity, one definition puts self-organising systems in between cosmos and chaos. According to that definition, systems that are neither totally ordered nor totally random seem to exhibit larger complexity, while perfectly ordered or perfectly random systems have very low structural complexity [Érdi 2008, 201-202; Grassberger 1986]. This parallels the phrasing that the most interesting developments happen "on the edge of chaos", that is, self-organising systems find their way between determined order and indeterminate disorder to exhibit a behaviour that is the most flexible, adaptable and creative [Kauffman 1993].

---

[c] I owe the term "suprasystem" to Arne Collen (personal communication).

In that same vein, the synergism hypothesis of Peter Corning [1983, 2003] postulates an evolutionary drift towards ever-increasing complexity because of the reinforcement of cooperative solutions by the synergy effects they produce. He lists three essential ingredients of complex phenomena:

(1) many parts,
(2) many relationships or interactions between them,
(3) and the production of synergy effects by them [Corning 1998].

Standard definitions of systems apparently do not reflect that demand. They concentrate on the components and the interaction of the components but leave open the question of modelling the most decisive feature of a complex system: its organisation, which is carried out through self-organisation.

An example is the definition of Hall and Fagen [1956, 18]:

> A system is a set of objects together with relationships between the objects and between their attributes.

This definition is problematic with regard to which objects belong to the system and which do not, if one attempts to distinguish between the system and its environment. Hall and Fagen define the environment [20] as

> the set of all objects a change in whose attributes affect the system and also those objects whose attributes are changed by the behavior of the system.

In later system definitions the environment is explicitly listed, but it still remains unclear which attribute or relationship makes an object an element of a system. In one interpretation, those objects that have closer or more direct relations or interactions with each other are regarded as elements of a system, whereas other objects belong in the environment.

According to the understanding presented here, the immediacy addressed above is due to self-organisation: those objects that take part in the self-organisation of that system are parts of the system. Self-organisation yields organisation. It therefore makes sense to include self-organisation in complex system definitions.

In his recent CESM model, Mario Bunge added a new constituent of the *definiens* of the system definition. He introduced the notion of processes in the definition of systems. He defines [2003, 35] a system s as the quadruple

$$\mu(s) = [C(s), E(s), S(s), M(s)]$$

whereby C stands for the composites, E for the environment, S for the structure and M for the "mechanism", that is, the processes, of the respective systems. Bunge thus highlights that a system cannot be defined solely by the set of elements and their relations to an environment. The processes that actually make the system a system must be included in the consideration. In complex real-world systems, this is, basically, self-organisation. This dynamic comes in a variety of manifestations depending on the specific nature of the system in question.[d]

Inspired by this definition, and considering that a system makes itself distinct from its environment by its very process of self-organisation (i.e. this is a secondary rather than primary feature of evolutionary systems), an ERD model is proposed according to the following definition of evolutionary systems:

Evolutionary System. *Evolutionary System s_e =def. a collection of*

(1) elements E that interact such that
(2) relations R emerge that – because of providing synergistic effects – dominate their interaction in
(3) a dynamics D.

This yields a distinction between micro-level (E) and macro-level (R) and a process (D) that links both levels in a feedback loop (Figure 4.9). Emergence and dominance are key properties of D. D is to ensure the perpetuation and elaboration of synergism.

---

[d] I hesitate to call this dynamic "mechanism" like Bunge does. This is because I want to stress the importance of avoiding notions that resemble mechanistic thinking. It was also not Bunge's intention to point out that the processes by which a system qualifies as system are mechanical in the sense of mechanistic thinking. Therefore I prefer to use the notion "dynamic", which meshes with the widespread notion of "dynamic systems" which defines complex systems.

4.2.4.2 *Emergent dynamics*

Emergence plays an undeniable role in EST. This gives self-organisation a touch of spontaneity, i.e. a touch of indeterminacy, because the order that is built up is not fully determined.

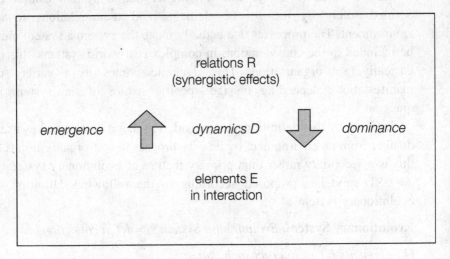

Figure 4.9. Evolutionary system.

In case of less-than-strict determinism and emergentism, the definition of causality, in terms of system-theoretical considerations, in contradistinction to the description of a mechanical universe but depicting complex systems, must be as follows [Hofkirchner 1998]:

(1) Inputs and outputs are not related in a way which can be plotted as bijective mapping. There are no transformation mechanisms which unambiguously turn the causes into the effects; causes and effects are coupled such that different causes can have the same effect, and the same cause different effects.

(2) Small changes in the causes may lead to considerable changes in the effects[e]. This is known as the so-called "butterfly effect"[f]. Dynamic systems exhibit sensitive dependence on initial conditions.

(3) The more complex a system, the less probable the return of a certain state in the future.

This is what ensues ontologically from findings in self-organisation research.

4.2.4.3 *Contingent trajectories*

The common feature of all non-mechanical causation is that the cause is an event which functions as a mere trigger of processes; these processes themselves depend on the nature of the system, at least inasmuch as they depend on the influence from the system's environment. The effect is an event in which this very self-organisation process finally ends up.

These connections reflect the fact that self-organising systems have the freedom to choose between several alternatives which make up a non-empty space of possibilities. This contrasts with mechanical systems, in which there is only one possibility (Figures 4.10 and 4.11).

---

[e] After A. Mittasch's book *Von der Chemie zur Philosophie*, Bertalanffy wrote about "instigation causality", by which "an energetically insignificant change" in an element of the system is "causing a considerable change in the total system" and is called a "trigger", in contradistinction to "conservation causality" [Bertalanffy 1950, 150].

[f] Originally, Edward Lorenz coined that term because the attractor of his computer model of the weather had the shape of a butterfly [Laszlo 1987]. The underlying point is that when entering parameters that differ in the decimal places, the computer programme would give quite different results. This gave rise to the metaphor that minute changes like the flap of a butterfly at one place on Earth could change the weather at some other place.

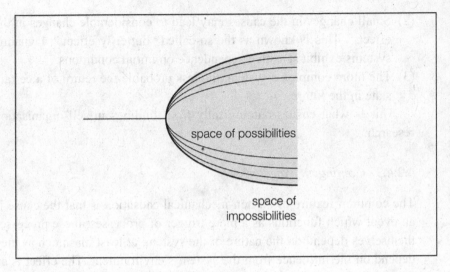

Figure 4.10. Trajectories in self-organising systems. Space of possibilities/impossibilities.

From this perspective, strict determinacy is merely a special case of causality. It applies if, and only if, the system is deprived of the freedom to choose between several alternatives and if the space of possibilities is narrowed down to one trajectory only.

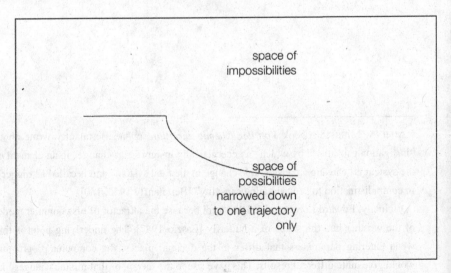

Figure 4.11. Trajectories in mechanical systems. Space of possibilities/impossibilities.

Propensive[g] isomorphies

The laws underlying different types of evolutionary systems are, as Bertalanffy pointed out, isomorphic, that is [Bertalanffy 1950, 138]

they hold generally for certain classes of complexes or systems, irrespective of the special kind of entities involved.

They are objective in the sense [Bertalanffy 1968, 87]

that the world (i.e. the total of observable phenomena) shows a structural uniformity, manifesting itself by isomorphic traces of order in its different levels or realms.

Bertalanffy was already clear about the fact that two types of laws belong to these isomorphies: those that can be formalised with mathematical rigor, and those that cannot be formulated in mathematical terms and thus exclusively by the use of ordinary language [Bertalanffy 1950, 137]. This makes it easy to connect isomorphies to the postnonclassical idea of chaosmic propensities and habits.

In this context, note that the German philosopher Herbert Hörz – as early as in 1970s – developed a philosophically based, system theoretical "integrated notion of law" [Schlemm 2003] that combines both the notion of dynamic law and the notion of statistical law [Hörz 1974, 365-366; 2009, 70-71]. Hörz uses the term "statistical" for his notion because the statistical aspect comprises the dynamic. He defines [Hörz 1982, 215]:

The philosophical conception of the statistical law regards laws (systems of laws) as general, necessary, and essential connections between objects and processes in a system, where, under the conditions of the system, a possibility is necessarily realized (dynamic aspect), but where there is a field of possibilities for the elements. A probability distribution exists for the random realization of this field (stochastic aspect) and the transition from one state into another is conditionally realized by chance with a certain degree of probability (probabilistic aspect).

The core idea is that the elements are not fully determined by the dynamics of the system [Schlemm 2003]. But an element might be a system itself, and a system might be an element of another system.

In analogy to this definition, the following definition of isomorphic laws based upon chaosmic propensities or habits is given. It uses the

---

[g] "Propensive" is the adjective form of the noun "propensity".

notion of system trajectories and uses the Popperian *dictum* that a law is a proscription rather than a prescription:

Systems law. *A systems law (an isomorphic law) exists if, and only if, given a class of evolutionary systems, there is a space made up of system trajectories impossible to realise for the systems such that*

(1) *the systems must realise by chance – i.e., without force from the environment – one possible trajectory out of a certain number of possible trajectories greater than one (dynamic aspect)*
(2) *for the realisation of all of which there exists a certain probability distribution (stochastic aspect)*
(3) *and for the realisation of each of which there exists a certain degree of probability (probabilistic aspect).*

This definition combines determinism with indeterminism. Bifurcation diagrams, which are an essential feature of phase spaces of evolutionary systems, always reveal both determinacy and indeterminacy, that is, mechanicity and spontaneity. Bifurcations mark possibilities for the system to go one way or another in building up its order. There is, however, no condition outside the system that compels the system to go this way or that way. It is therefore up to the system itself. What is determined is that the system has to go one way or another, but it is not determined which way it will go.

Subjective agency

The system itself shows an activity that selects one of the several possible ways of reacting. George Kampis [1991, 257-258] clearly points to the property that causation has if emergence and agents are implied, although spontaneity does not make it a supernatural or amaterial phenomenon.

Material causation is just a word. It is clear ... that what it means is that *we do not know anything about the causes* that determine things by an invisible and unapproachable 'actor'. The actions of this actor bring forth something new. It has to be taken very seriously that molecules, thoughts, artefacts, and other qualities never existed in the Universe before they were first produced by a material causation. Therefore, irreducible material causation is creation per se: *free construction of new existence, with new properties, i.e., that in no conceivable form pre-exist either physically or logically.* Before they are already there, absolutely no hint can be gained about their possibility, about their properties, about how they come into being, what they will look like and what will happen to them next...

Robert Rosen [1986] mapped Aristotelian causes to modern mathematical objects that affiliate with systems in the following way:
- material cause associates with initial values;
- formal cause associates with parameters;
- efficient cause associates with driving force.

Just final cause remains unassociated. It is proposed here to associate it with attractors [Weisstein n.d.]:

> An attractor is a set of states (points in the phase space), invariant under the dynamics, towards which neighboring states in a given basin of attraction asymptotically approach in the course of dynamic evolution.

Returning to the philosophical notion of subjectiveness (Figure 4.4), one can ascribe different systemic qualities to the subjects identified; in system theoretical terms, these subjects turn into agents that demonstrate agency (Figure 4.12).

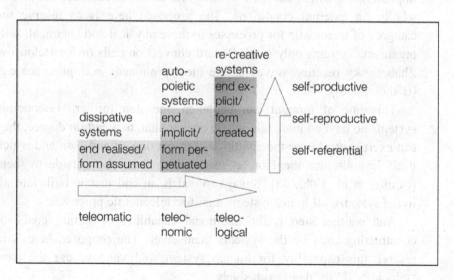

Figure 4.12. Agents.

Attractors

As to the horizontal axis – the axis of end-directedness (Figure 4.4) – it is helpful to use (and to slightly modify) a well-known distinction Ernst Mayr introduced into the theory of biology [1974]: he distinguished between "teleomatic", "teleonomic", and "teleological" processes. The first evoke an analogy to automatic, and the second an analogy to economic processes. According to Mayr, teleomatic processes end up in an end as a consequence of physical laws such as gravity, entropy decay, reaction gradients. Processes are teleonomic due to an in-built programme which directs them towards an end like in homeostasis, ontogeny, biotic reproduction. Teleological processes are present with the intervention of cognitive mechanisms, mostly human.

With Mayr, teleomatic processes are strictly mechanical: they can be described and explained in terms of strict determinism. But the new paradigm reveals systems more interesting than purely mechanical systems, namely self-organising systems, however primitive they may be. The latter have an end to which these systems tend. This end is implicit and internal, but the conditions for its satisfaction depend almost wholly on external conditions. The proposed here is to reserve the category of teleomatic for processes in these physical and chemical, self-organising systems only (with Bénard convection cells or the Belousov-Zhabotinsky reaction waves as the most prominent examples, see e.g. [Bishop 2008]).

This line of thought can be taken one step further. Teleonomic systems go beyond mere teleomatic ones in that, to a certain degree, they can exert control over the conditions for meeting an end – an end which itself is built into them or, at least, given from the outside to them [Coulter et al. 1982, 43]. Since survival is an end that is built into all living systems, all living systems manifest teleonomic processes.

And another step is the additional capability of setting goals, of constructing ends by the systems themselves. The proposed here is to reserve this capability for human systems only and to use the term "teleological" for them exclusively.

This approach yields a clear-cut logical distinction (Figure 4.12) between

(1) systems and processes that result in an end when conditions are met – termed teleomatic,
(2) systems and processes that have an in-built end and can control the conditions to meet it – termed teleonomic, and,
(3) systems and processes that construe a diversity of ends – termed teleological.

Downward causation

The second axis – the axis of formative power (Figure 4.4) – also merits closer examination. Here similar steps can be introduced. There are systems and processes that manifest patterns. Pattern is form, that is, a superstructure that refers to a basis that refers to the superstructure, and so on. These are macro- and micro-levels that co-exist and influence each other; this influence is more important than that from outside. The system is produced by its elements, and the system constrains and enables its elements at the same time. As this works through dissipation of entropy, Ilya Prigogine [1980] called the emerging structures "dissipative". The fluid particle in the Bénard convection cell is prompted to contribute to the cell structure, which emerges from the activities of all particles.

The above systems are accompanied by systems and processes that maintain the form they show, i.e. they maintain a stable form while matter is changing. This is the case with all living systems. Maturana and Varela [1980] designated them with the neologism "autopoietic" in order to denote the fact that they are systems that produce themselves – in an interpretation that is favoured in the context of EST – by constraining and enabling their elements to produce new elements that (re-)produce the (form of the) systems.

Finally, there is a third set of systems and processes that change their form in a deliberative way: they transcend themselves, invent themselves, create themselves. This is what Austrian philosopher Erich Jantsch [1987] meant when referring to "re-creative" systems at the human level. Cases in point are different societal formations (e.g. capitalism, communism), different party programmes transformed into government plans transforming society, individual life planning and different projects individuals carry out in order to become what they want to be.

The proposed here (Figure 4.12) is to attribute to them the adjectives of being

(1) self-referential[h] ("reference" not referring to propositions but to concrete, real-world systems establishing circular causality between parts and whole),

(2) self-reproductive ("reproduction" not in the traditional biological sense but in the system theoretical sense of producing the conditions the system requires to maintain itself as usual in social sciences such as political economy which gives the biological notion a new drive), and

(3) self-productive ("productivity" going beyond the Maturana-Varela "poiesis" – which refers to "techné" – in that these systems really produce themselves anew in the course of praxis – which refers to "phronesis").

Dissipation, autopoiesis, re-creation

This perspective reveals

(1) that all self-organising systems can be considered teleomatic regarding the end-directedness and self-referential regarding the formative power,

(2) that all biotic self-organising systems are that subset of self-organising systems that are, in addition, teleonomic regarding the end-directedness and self-reproductive regarding the formative power,

(3) and that all social systems are that subset of biotic self-organising systems that are, in addition, teleological regarding the end-directedness and self-productive regarding the formative power.

This reveals degrees of subjectiveness and thus agency in evolutionary systems. We can now identify proto-subjects as agents that are simple dissipative systems and quasi-subjects as agents that are

---

[h] See John Mingers [1997], who develops a seven-class typology of self-referential systems. His classification resembles the evolutionary trend I propose here from self-referentiality to self-reproductivity to self-productivity.

simple autopoietic systems while reserving the property of being a subject in the full sense of the word for agents that are re-creative systems only.

Note, anyway, that there is a continuum in the evolution of agents.

4.2.4.4 *Path-dependent nestedness*

Self-organisation may be viewed as the way evolutionary systems come into existence or change their structure, state or behaviour and the way they maintain themselves (their structure, state or behaviour).

Self-organisation has diachronous as well as synchronous aspects. In the literature they are usually referred to as separate specification and scalar hierarchies among others [Salthe 1996]. However, it can be shown that both hierarchies under certain circumstances are only two sides of the same coin. A stage concept of systemic evolution seems able to reconcile them.

The core of EST is such a stage model. It is a phase model and a layer model in one. The stage model of evolutionary systems is based upon the principle of emergentism and the principle of asymmetrism. Emergence takes place in transitions in which systems are produced by the interaction of proto-elements. Asymmetry describes the suprasystem hierarchies in which subsystems are encapsulated.

The shift from one phase to a subsequent phase is therefore a shift onto a new layer. The new system includes this additional layer. It encapsulates what previously were autonomous systems as subsystems and shapes them to reflect the dominance relation. However, the newly formed system will always depend on the functioning of its subsystems. When they cease to support the system, it will break down.

Metasystem transition

The first shift which describes the genesis of a system is known as a "metasystem transition" [Turchin et al. 1999]. It concerns the diachronous aspect of the stage model and is about the evolutionary dynamics of systems. The logic by which the metasystem transition is reconstructed assumes the following phases:

(1) In a first phase there is only a multitude of entities, which later on will become elements of the system to be formed. In this phase they cannot be addressed as elements because there is no system yet. They have no bindings to each other. This phase may be called individual phase (Figure 4.13).

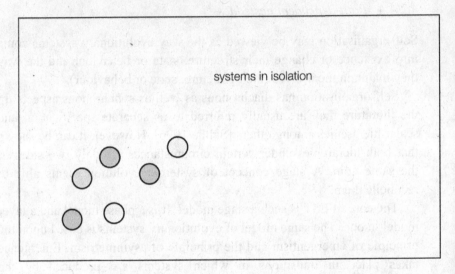

systems in isolation

Figure 4.13. Individual phase of metasystem transition.

(2) In the second phase these entities begin to develop relations among themselves: they interact with each other. But this interactive relationship need not be durable or stable, and can vanish according to the changing activities of the entities involved. In this interactional phase, processes may still be reversible (Figure 4.14).

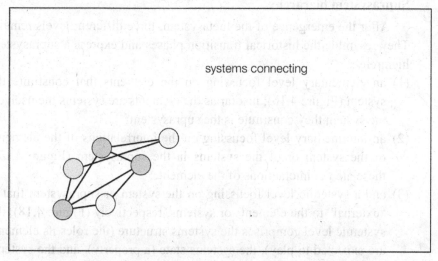

Figure 4.14. Interactional phase of metasystem transition.

(3) In a third phase, the interaction produces a system. Durable, stable relations are established among the entities, which by then become elements of solely that system. This integration phase makes the changes irreversible. A new system has emerged (Figure 4.15).

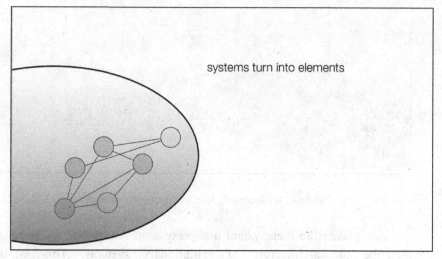

Figure 4.15. Integrative phase of metasystem transition.

Suprasystem hierarchy

After the emergence of the metasystem, three different levels remain. They resemble the historical transition phases and express a suprasystem hierarchy:

(1) an elementary level focussing on the elements that constitute the system (Figure 4.16); insofar as the elements are systems themselves, the system they constitute is the suprasystem;

(2) an intermediary level focussing on the interrelations of the elements of the system or of the systems in the suprasystem (Figure 4.17); these are the interactions of the elements;

(3) and a systemic level focussing on the system or suprasystem that is "external" to the elements or systems, respectively (Figure 4.18); the systemic level comprises the systems structure (the roles its elements are expected to play), the systems state (a property), and the systems behaviour (exhibited vis-à-vis the environment).

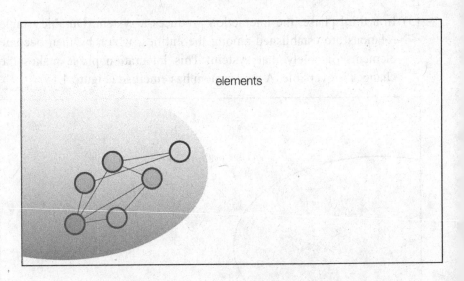

elements

Figure 4.16. Elementary level of suprasystem hierarchy.

This makes the metasystem a suprasystem. Altogether, these levels describe the architecture of evolutionary systems. This is the synchronous aspect.

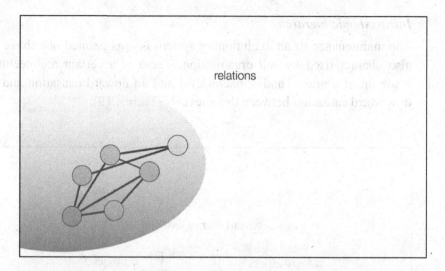

Figure 4.17. Intermediary level of suprasystem hierarchy.

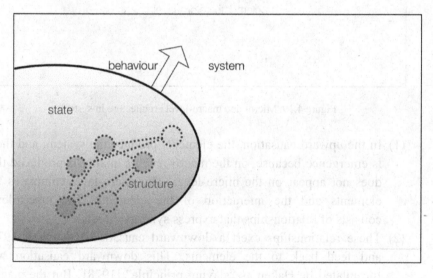

Figure 4.18. Systemic level of suprasystem hierarchy.

These suprasystemic levels apply to two different contexts, one intrasystemic and one intersystemic.

Intrasystemic hierarchy

The maintenance of an evolutionary system is – as pointed out above – also characterised by self-organisation. There is a certain architecture made up of a micro- and a macro-level and an upward causation and a downward causation between these levels (Figure 4.19).

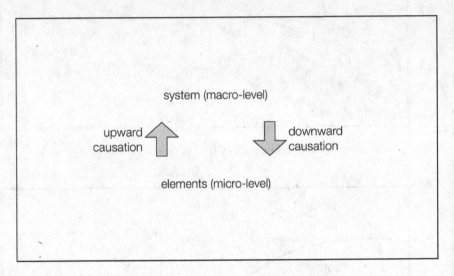

Figure 4.19. Micro- and macro-level architecture in systems.

(1) In the upward causation, the elements produce the system, and there is emergence because, on the macro-level, a quality is produced that does not appear on the micro-level. The micro-level comprises the elements and the interaction of the elements. The macro-level consists of relationships that express synergy effects.

(2) These relationships exert a downward causation [Campbell 1974] and feed back to the elements. This downward causation was formulated by Haken as "slaving principle" [1978][i]. But the macro-level functions not only as a constraint but also as an enablement for the agency of the elements, as pointed out in the Principle of Unity-

---

[i] Haken was fond of recounting that social scientists criticised his anthropomorphic metaphor. He preferred then to depict it as a dialectic process.

Through-Diversity. The characterisation of the downward causation as "constraining" and "enabling" in one goes back to British sociologist Anthony Giddens' theory of structuration [1984].

Elements and system work as parts and whole. Bertalanffy, for example, took Nicholas of Cusa's idea *"ex omnibus partibus relucet totum"* ("each part reflects the whole") as a point of departure. As early as in 1928, Bertalanffy wrote with regard to the organism [Bertalanffy 1928, translation into English quoted after Pouvreau and Drack 2007, 305]:

> The characteristic of the organism is first that it is more than the sum of its parts and second that the single processes are ordered for the maintenance of the whole.

Here he anticipated Haken's slaving principle for the organic world (parameters that change more slowly are those that enslave the rest of the parameters). With his empirical findings he laid the foundation for what Maturana and Varela later called autopoiesis (the system is a network of elements that produce new elements that maintain the network) [Bertalanffy 1932, translation into English quoted after Pouvreau and Drack 2007, 309]: Bertalanffy discovered that there is

> maintenance of the organized system in a dynamical pseudo-equilibrium through the change of its components.

When characterising this intrasystemic hierarchy, Bertalanffy asserted [Bertalanffy 1950, 135]

> the necessity of investigating not only parts but also the relations of organization resulting from a dynamic interaction and manifesting themselves by the difference in behavior of parts in isolation and in the whole organism.

Note that he distinguishes not only between the level of parts and the level of the whole, but also between the dynamic interaction of the parts and the relations of organisation. He clearly locates the interaction on the parts' level and the relations on the whole's level. And he considers the following relationship between the interaction and the relations: the relations, on the one hand, result from the interaction and, on the other, are manifest in the behaviour of the parts in that the behaviour is different from the behaviour when in isolation. It therefore follows that there are two processes in systems:

(1) one bottom-up in which interactions on the level of the parts result in relations on the level of the whole,

(2) and one top-down in which relations on the level of the whole manifest themselves on the level of the parts, viz., in their behaviour.

In summary, the maintenance of a system functions such that the system (via downward causation) makes its elements produce (elements that produce) the system itself (via upward causation). That is, the system (the self) refers to itself, albeit by referring to its elements. There is self-reference in each self-organising system.

Intersystemic hierarchy

The suprasystem hierarchy signifies not only the internal architecture of any system but also, as an outcome of evolution, the external architecture of systems.

Bertalanffy arrived at the conclusion [1950, 164]:

> Reality, in the modern conception, appears as a tremendous hierarchical order of organized entities, leading, in a superposition of many levels, from physical and chemical to biological and sociological systems.

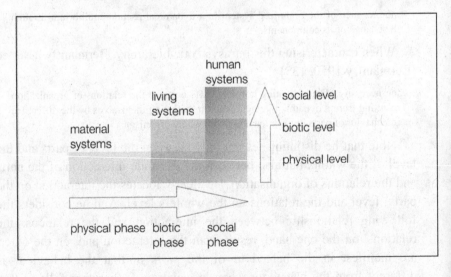

Figure 4.20. Real-world holarchy.

"Speaking in the way of gross oversimplification", he conceived of three levels: "physical nature; organisms; and human behavior, individual and social" (Figure 4.20). This, in general, concurs with the differentiation given in Figure 4.12.

And here [Bertalanffy 1959, 67]

the notion of emergence is essentially correct: each higher level presents new features that surpass those of the lower levels

and cannot be reduced to those of the lower levels [Bertalanffy 1950, 165]:

When emphasizing general structural isomorphies of different levels, it asserts, at the same time, their autonomy and possession of specific laws.

In 1967, Arthur Koestler coined the terms "holarchy" and "holon". This was a remarkable contribution to conceiving the synchronous aspect of self-organisation regarding the totality of real-world systems. "Holarchy" denotes the hierarchy of "holons". A holon is a self-contained whole made up of its subordinate parts, while, in return, it is itself a part dependent on another whole [Koestler 1967].

Combining the above, the ontology of evolutionary systems can, finally, be defined:

Evolutionary Systems Ontology (Dynamics and Architecture). *Evolutionary Systems Ontology is that view of the interconnectedness of real-world events or entities that applies the Stage-Model Principle. The Stage-Model Principle is based upon the Principle of Less-than-Strict Determinacy as well as upon the Principle of Self-Organisation.*

The **Stage-Model Principle** states: The view of the interconnectedness of real-world events or entities is to model self-organising systems and has to acknowledge irreversibility of systemic evolution and irreducibility of evolutionary systems. Nonetheless, that view is able to model (I) the process of self-organisation as well as (II) the self-organised structure of systems.

In particular, the Stage-Model Principle consists of two principles:

(I)   The **Principle of Historicity** states with regard to diachrony: Model the path dependency of self-organising systems such that one step taken by a system in question – that produces a layer – depends on the step taken prior to that but cannot be reversed!

(II)   The **Principle of Holarchy** states with regard to synchrony: Model
         the internal and external hierarchical order of self-organising
         systems such that layers – that are produced by steps – build upon
         layers below them but cannot be reduced to them!

Evolutionary Systems Ontology (Dynamics and Architecture) rejects
opposing mechanicity to spontaneity. Both are justified within certain
boundaries, and reality has to be modelled accordingly.

Chapter 5

The Adjacent Necessary[a]

> The first matter of importance is to note that, from the standpoint of the formalisms being compared, *the encoding and decoding arrows* [...] [of a Modeling Relation between a Natural System F1 and a Formal System F2 – W.H.] *are unentailed.* In fact, they belong to neither formalism, and hence, cannot be entailed by anything in the formalisms. The comparison of the two inferential structures, like F1 and F2, thus inherently involves something outside the formalisms, in effect, a *creative* act, resulting in a new kind of formal object, namely, the modeling relation itself. It involves art.
>
> – Robert Rosen: Life Itself, 1991 –

Completing the praxiological and ontological considerations of the new *weltanschauung* calls for examining the epistemological implications of, first, POE and, second, EST. The first serves as a fundament for the second. The implications of POE provide a fresh perspective on comprehension, that is, explanation and understanding, those of EST deliberate on systems analysis and synthesis of systems, i.e, on Evolutionary Systems Methodology. Principles for comprehension and Evolutionary Systems Methodology represent another two steps towards a UTI.

5.1 A Fresh Perspective on Comprehension

Table 2.2 underlines that the basic question of epistemology in POE is "explainability vs. unexplainability". Table 2.3 lists the four possible answers according to the way of thinking applied:

---

[a] My wording is inspired by Stuart Kauffman's theory of the "adjacent possible". In a way, my argument is converse.

(1) The reductionistic answer is scientism.
(2) The projectivistic answer is anthroposociomorphism. Both answers extol rationalistic explainability.
(3) The disjunctivistic answer makes its irrationalistic disbelief in explainability a belief in unexplainability as a matter of principle.
(4) The integrativistic answer goes beyond the rationalism/irrationalism divide. Reflexive rationalism favours a fresh perspective on explainability.

The following describes each of the above answers in terms of basic assumptions, in terms of tools that are used, in terms of conditions that are taken into consideration and in terms of how essence and phenomena are related (Table 5.1).

Table 5.1. Four research programmes from the philosophical point of view.

	basic assumption	tools used	conditions considered	essence and phenomena correlated
scientism	naturalism	analysis		phenomena are reducible to an essence
antropo-socio-morphism		synthesis	sufficient condition	phenomena are projectable onto an essence
irrationalism	culturalism	anything goes	no condition	no match
reflexive rationalism	incomplete deducibility	ascendence from the abstract to the concrete	necessary but not always sufficient condition	unity to-be-produced

5.1.1 *Scientism*

The scientistic programme suits mechanicist determinism and modernism. It supposes a positive answer when asked whether everything is, in principle, explainable (Table 5.1, line 1). It is

Naturalism. Explanations are sought with regard to nature – natural processes and natural structures. The possibility of supernaturally caused events or entities is ruled out. Hence the name "naturalism". Another connotation is insinuated when it comes to explanations of social events or entities by resorting to nature. This is a special case of naturalism – naturalism in the narrower sense.

Naturalism in the broader sense includes not only the mainstream natural science approach but also the mainstream engineering science (technological science) – "technicism" – and mainstream formal science (logics, mathematics) – "formalistics" – approaches as far as they make use of the same tool:

Analysis. The predominant feature of occidental science has been to cut the object of inquiry in pieces, to separate them and to single out those pieces that are supposed to count for the entire object. Given such a piece, one is supposedly in a position to explain and predict the whole event or entity. Analysis plays the role of a

Sufficient condition. A sufficient condition p for something that is conditioned q can be formalised as antecedent of an implication $p \rightarrow q$. p being the case suffices for q being the case. Every time p is true, q must be true too. The truth of p is transferred to q.

The implication allows for applying the *modus ponens*. The *modus ponens* is paradigmatic for deductive thinking. A deduction is a conclusion from premises in which the truth from the premises is transferred to the conclusion. Given the implication $p \rightarrow q$ as premise 1 and given p as premise 2, conclusion 3 follows from the premises: q.

This line of reasoning is compelling.

Explanations are likened to this kind of reasoning. Speaking in terms of formal logic, an explanation or prediction is the deduction of a conclusion from premises such that the conclusion describes what is to be explained or predicted, and such that the premises are made up of descriptions of what together is expected to do the explaining or predicting. The conclusion is termed *explanandum*, the premises are termed *explanans* (Figure 5.1).

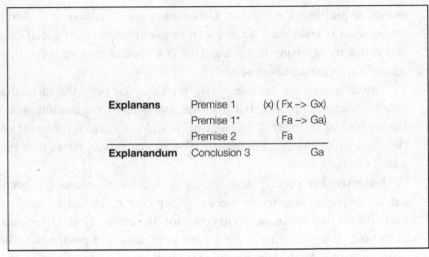

Explanans	Premise 1	(x) (Fx –> Gx)
	Premise 1*	(Fa –> Ga)
	Premise 2	Fa
Explanandum	Conclusion 3	Ga

Figure 5.1. The basic scheme of deductivism after Popper, as well as Hempel and Oppenheim.

After Hempel and Oppenheim [1948], this scheme is called deductive-nomological if it couples empirical and theoretical knowledge by subsuming facts (empirical) under some law (theoretical) that covers those facts. The explanation of a phenomenon is translated into the answer to the question [1948, 136]: "according to what general laws, and by virtue of what antecedent conditions does the phenomenon occur?"

After Popper's *The Logic of Scientific Discovery*, the original German version [1935] of which is cited by Hempel and Oppenheim, the general law assumes, in terms of predicate logic, the form of a universal implication (Figure 5.1).

Premise 1 states for every individual x that if x has the property F, then x also has the property G.

Premise 1* is an instantiation of the universal implication for the individual a (this premise is listed for reasons of completeness of the whole argument and exactness of the intended deduction but is no independent premise).

Premise 2 consists of a sentence that reflects the empirical observation that the individual a has the property F.

Because premise 2 is an instantiation of the if-component of the universal implication, the application of the rule of *modus ponens* derives the instantiation of the then-component as conclusion 3, which is a sentence that states that a has the property G. This sentence represents solely that phenomenon which was observed and should be explained or will be observable if it is predicted.

The conclusion must be realised when the premises are the case. Thus premises 1 and 2 suffice to explain or predict what is stated in the conclusion. Together they describe the sufficient condition. This means there is a clearly defined one-way road in the logical argument from the premises to the conclusion. The conclusion is deduced from the premises or, more succinctly, the conclusion is reduced to the premises. The search for explanation is the search for the sufficient condition, while prediction is the conviction of possessing knowledge of the sufficient condition.

The belief in complete deducibility of propositions representing phenomena to be explained or predicted is known as "deductivism".

Propositions from which those propositions representing phenomena are deduced (traits to which phenomena are reduced) are deemed essential. Accordingly, deductivism produces an

Essence short cut. Neither the universality of the universal implication nor the necessity of the implication can guarantee that the explanation (prediction) reveals an essential trait of the phenomenon. What is universal, that is, common to phenomena, needs to be necessarily in common to be considered essential. Popper's example of a universal implication "All swans are white" (for everything that is a swan it is implied that it is white) clearly fails to point out an essential feature. Analysis abstracts a common feature from the concrete phenomenon and holds it to be essential. As a result, the essence is simplified to meet the phenomenon at the same level. However, commonality is only an indication, if any, for the essential.

In the above approach, nothing new and nothing that goes beyond the parts and forms a whole can be explained (predicted). This is because the knowledge of the *explanandum* can be inferred from, and is implied by, the knowledge of the *explanans* and thus cannot contain a piece of knowledge that is not already contained in the knowledge of the *explanans*. The phenomena of novelty and wholeness, however, can be comprehended only by incorporating knowledge that exceeds the knowledge of what is old compared with novelty and what are parts in relation to wholeness.

5.1.2 *Anthroposociomorphism*[b]

Like scientism, anthroposociomorphism believes in complete deducibility as a means of explanation and prediction. The difference is merely the culturalistic viewpoint, the synthetic bias, and the projective attitude when essence comes into sight (Table 5.1, line 2).

Culturalism. In contradistinction to naturalism, the premises that are sought for explanation or used for prediction refer to the social, human and societal – socio-economic, socio-political, socio-cultural – domain. Social causes are addressed in the *explanans*. The *explanandum* might cover social (individual, societal) phenomena or natural ones. In the latter case it becomes clear why the term "anthroposociomorphism" is used: natural phenomena are explained (predicted) in the way social ones are explained (predicted).

Synthesis. In order to arrive by deductive reasoning at the desired conclusion, concrete features of the phenomenon in question are collected and added to the essence (clearly a kind of inductive procedure during the course of explanation). This involves a projection of the unique onto the universal. Once the knowledge of the concrete is contained by the knowledge of the *explanans*, the knowledge of the *explanandum* can be derived.

Phenomena overdetermination. In naturalism, phenomena tend to be underdetermined by the abstract essence and become abstract as well. In anthroposociomorphism the opposite is the case. Phenomena tend to be overdetermined by an essence that contains the concrete.

---

[b] I include the syllable "socio" in "anthropomorphism" (in analogy to the term "anthroposociogenesis") to highlight that not only the human individual but also human society is taken as a point of departure.

5.1.3 *Irrationalism*

Both naturalism and anthroposociomorphism rely on deductivism, which makes them rationalistic. Rationalistic culturalism, however, has an irrationalistic culturalistic twin that contends the unexplainability and unpredictability of cultural (social) events and entities. This is based on the supposed inability to deduce explanations, because of alleged nondeducibility (Table 5.1, line 3).

Anything goes. Instead of explanations via abstraction or concretisation, via analysis or synthesis, this approach favours a different way of understanding (German "Verstehen"), which is a central term in the tradition of phenomenology and hermeneutics. Here, sectors of reality that cannot be explained are described and interpreted according to some sort of sense. This sense can virtually be of any type. Hence, the tool used can be likened to the motto of Paul Feyerabend's methodological anarchism "anything goes" [Feyerabend 1975]. This touch of arbitrariness leaves this two-cultures thinking open to criticism for lacking of rational substantiation of its background ideas.

Neither sufficient nor necessary condition. The programme of "Verstehen" offers a quite different option and postulates a quite different approach of comprehension as opposed to the "nomothetic" way: an "idiographic" way. Windelband [1894, translated by Mos 1998] closed his famous lecture *History and Sciences* with the words:

> The law and the event remain to exist alongside one another as the final, incommensurable forms of our notions about the world.

This means that there is no need for premises to deduce conclusions. Everything that is needed is descriptions of events and entities, but no law is required for an explanation.

No interrelation of essence and phenomena. Since any interpretation may work, there is no inherent relation between the essence and the phenomenon.

5.1.4 *Reflexive Rationalism*

Deductive rationalism and anti-deductive irrationalism prolong the trap of being locked in either the programme of explanation and prediction, on the one hand, or the programme of interpretation in the sense of "Verstehen", on the other. A fresh look is needed to get out of the trap (Table 5.1, line 4).

Incomplete deducibility. In contrast to the view imposed by naturalism, it is not unscientific to get by without experimental or mathematical methods, and, in contrast to the anthroposociomorphistic globalisation of cultural thinking, explanations of natural science are not merely a misunderstood variety of traditional understanding in the humanities. We can even go one step further. Deductive rationalism holds that nondeductive reasoning is not scientific at all. This, however, is not true. Nondeductive reasoning is merely nonreductive and nonprojective, but need not be disjunctive in making things that need to be understood unrelated to any conditions. In contrast to the disjunctive culturalism, it is not sensible to divide the applicability of scientific methods along the line dictated by the differentiation of nomothetic and idiographic. Both naturalistic and culturalistic philosophies deal with the description of events and entities and the comprehension of their becoming and being, be this in the form of explanation, prediction or understanding.

Ascendence from the abstract to the concrete. Naturalism is content with reducing the concrete to something abstract (analysis), and rationalistic culturalism settles for projecting something concrete to the abstract (synthesis). Comprehension, however, needs a combination of analytical and synthetic tools: here, analysis decomposes the real, concrete something as it can be observed (in an undifferentiated way), and synthesis recomposes the concrete from the differentiated abstract items in the way they are composed in reality.

The reconstruction of the concrete by the human mind and science presupposes abstraction but must not stop there. Abstraction is simply a means for reconstruction. Concretisation is necessary for complementing the task, though not for completing the task. Concretisation is a step-by-step process in which the finding at each step is enriched by something that is the result of abstraction and is abstract itself. But when combined, they form an ever new concrete. Thus, synthesis integrated with analysis is an ascendence from the abstract to the concrete. Each step is an emergent one. From one step to the next there is no deduction.

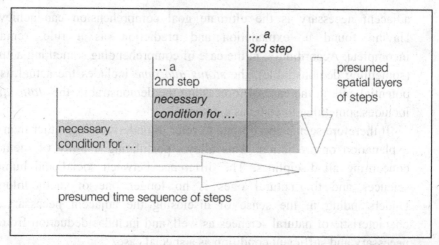

Figure 5.2. Presumed steps are linked by providing necessary conditions for the subsequent ones.

The adjacent necessary. Each step is a necessary condition for the next step (Figure 5.2): without a step as a point of departure, there can be no next step. But, given a certain step, the next step does not need to occur at all, and if it occurs, it does not need to occur in the specific way it does and it does not need to occur at a specific point of time. Thus it cannot be deduced.

Since comprehension is not a deduction but an ascendence from one necessary condition to the next, it is a search for the necessary condition that made possible what is to be comprehended. And this necessary condition should be as close as epistemic circumstances allow. This makes comprehension the search for the epistemologically adjacent necessary.

This adjacent necessary is historical in character because it depends on the historically given circumstances of how detailed inquiry can be. In the course of scientific progress, the adjacent necessary will get closer and closer.

Only in the case of mechanical systems can the historical process of comprehension come close enough to the object of comprehension to make the necessary condition a sufficient one and yield a deduction. Deduction is then a special case of comprehension. In all other cases the adjacent necessary is the ultimate goal comprehension can achieve. Having found it, explanation and prediction, as a rule, remain incomplete. Accordingly, in the case of comprehending something actual (something demonstrable), the *status quo ante* includes the actual as a potential and, in the case of forecasting the demonstrable, the *status quo* includes something to come as a potential.

It therefore seems appropriate to refer to understanding rather than to explanation or prediction. This allows postulating a unity of method concerning all disciplines. The difference between social and human sciences and the natural ones is no longer one of methodology. Understanding in the sense of disclosing the adjacent necessary is characteristic of natural sciences as well, and includes deduction from a necessary and sufficient condition as a special case.

The leitmotif of the old approach is looking for the "necessary and sufficient condition" or looking for anything else because there is "no condition at all". In contrast, the new guiding idea of epistemology is the search for the "necessary, but not in every case sufficient condition".

Unity of essence and phenomena. The ascendence from the abstract to the concrete is realised via the ascendence from the necessary to the conditional and from the universal to the particular, by which a reconciliation of essence and appearance is intended.

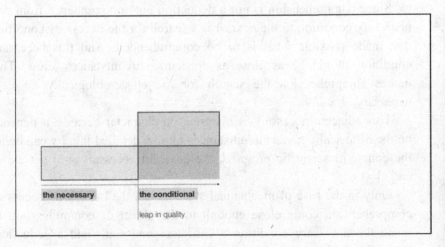

Figure 5.3. The ascendence from the necessary to the enabled conditional.

Like the terms "abstract" and "concrete", the "necessary" and the "conditional" as well as the "universal" and the "particular" are epistemological, not ontological concepts. The "necessary" describes the necessary condition and the "conditional" describes that which is conditioned. The "universal" denotes what comprehension finds in common and the "particular" denotes what it considers distinctive. Abstract is any description that separates the necessary from the conditional or the universal from the particular.

In our multistage scheme the necessary prepares the ground for another step that is possible and contingent on the necessary – the enabled conditional (Figure 5.3).

At the same time there is another leap in quality which leads from that which all steps so far have in common – the universal – to the particular (that which is characteristic solely of the particular step in question) (Figure 5.4).

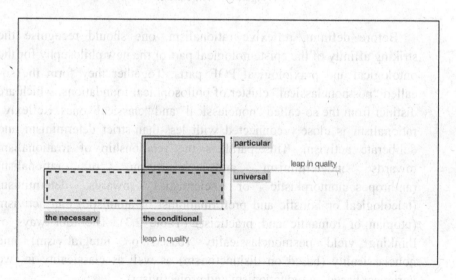

Figure 5.4. The ascendence from the universal to the particular.

The concrete can then be conceived of as the conditioned particular together with its conditioning universal on which it rests (Figure 5.5). Altogether, an ascendence from the abstract to the concrete is achieved.

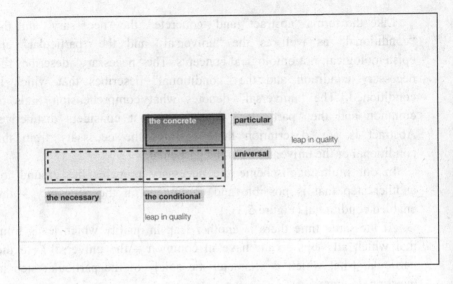

Figure 5.5. The concrete.

Before defining reflexive rationalism, one should recognise the striking affinity of the epistemological part of the new philosophy for the ontological and praxiological POE parts. Together they form the so-called "postnonclassical" cluster of philosophical foundations, which are distinct from the so-called "nonclassical" and "classical" ones. Reflexive rationalism is closely connected with less-than-strict determinism and deliberate activism. This parallels the relationship of irrationalism towards indeterminism and inactivism; of rationalism (anthroposociomorphistic or scientistic) towards determinism (teleological or holistic and preformationist or atomistic) and activism (utopian or romantic and practicist) (Table 2.3). The four ways of thinking yield postnonclassicality (based on integrativism) and nonclassicality (based on disjunctivism) as well as classicality in two varieties (based on reductionism and projectivism).

This calls for distinguishing between objects of praxis (the praxic dimension), objects of reality (the ontic dimension) and objects of method (the epistemic dimension) [Hofkirchner 2004, 2011b]. Objects of praxis O_p are those acted upon. Objects of reality O_o are those existing as such. And objects of method O_e are those in our heads. According to the way we (assume to) act on objects O_p, we assume how they exist independently of our actions as O_o. And according to the way (we assume) the objects O_o exist, we assume methods of investigation and representation by which we manipulate the objects O_e in our heads. And according to the way we (assume to) link objects O_p in praxis, (we assume) they are able to be linked as objects O_o in reality: it is according to the latter that (we assume) they have to be linked by our method as objects O_e.

Let O_x^{t1} and O_x^{t2} indicate the same object O_x at the point of time t_1 and t_2, respectively, whereby $x = \{p, o, e\}$ indicating the dimension, and let the arrow $\rightarrow$ indicate an unambiguous transformation while the sign $\nearrow$ signifies a transformation that involves an ineluctable ambiguity.

We can then describe the clusters from which the "postnonclassical" cluster differs as follows.

First, in the case of the reductionist, mechanistic variety of the "classical" cluster, we have:

(1) on the praxic level $O_p^{t1} \rightarrow O_p^{t2}$; that is, the action applicable is a "brute force" operation which leads unambiguously from the object in an initial state to the object in a well-determined final state; humans can apply this operation only when functionalising cause-effect-relationships that rest upon the ontic level;

(2) thus, on the ontic level we have $O_o^{t1} \rightarrow O_o^{t2}$; that is, the object at t_1 is causally transformable into the object at t_2 by pure necessity; this is the case with strict determinism; objects transform in this manner only when embodying on the ontic level deductive logic or computable functions or algorithmic prescriptions that are located on the epistemic level;

(3) thus, on the epistemic level we have $O_e^{t1} \rightarrow O_e^{t2}$; that is, the object at time t_2 is derivable from the object at time t_1, given particular conditions; the outcome is necessitated in a compelling way: a conclusion is unavoidably drawn from premises in an inference; a mathematical solution unavoidably results from inputs in formulae; and data is unavoidably processed by algorithmic computer programmes.

Second, in the case of the projectivist variety, there is a castling of O_x^{t1} and O_x^{t2}:

(1) on the praxic level there is $O_p^{t2} \to O_p^{t1}$, which is a kind of utopian or romantic back-propagation from the final state;

(2) this necessitates on the ontic level $O_o^{t2} \to O_o^{t1}$, which is a supposed functionalisation of final cause working with the same inevitability as strict determinism works in reductionism;

(3) and this necessitates on the epistemic level $O_e^{t2} \to O_e^{t1}$, which is a deduction as well.

Third, in the case of nonclassicality, there is no feasible connection between O_x^{t1} and O_x^{t2} at all.

Finally, we arrive at the formal description of postnonclassicality. It is only in this case that transformations from O_x^{t1} to O_x^{t2} are open for openness so as to yield:

(1) on the praxic level according to the Principle of Limited Controllability $O_p^{t1} \underline{\uparrow} O_p^{t2}$; the outcome cannot be controlled entirely, and there is only a certain bandwidth of possible outcomes that can be controlled;

(2) on the ontic level according to the Principle of Less-than-strict Determinacy $O_o^{t1} \underline{\uparrow} O_o^{t2}$; causal relations inhere uncertainties and thus do not "obey natural laws" as viewed in a mechanistic concept, but are rather inclined to propensities our cosmos is displaying;

(3) on the epistemic level according to what can be formulated as Principle of Incomplete Explainability/Predictability (see below) $O_e^{t1} \underline{\uparrow} O_e^{t2}$; the base from which the transformation on the ontic level starts has to be codified on the epistemic level as a necessary condition only (but not as a sufficient one) in order to do justice to the emergent character of the "result" of the transformation, which represents a new quality; dialectic logic with its sublation scheme is a good candidate for grasping this relationship.

This postnonclassical mode of comprehension can be defined as a type of rationalism:

Reflexive rationalism. *Reflexive rationalism is that programme of comprehension that applies the Principle of Incomplete Explainability/ Predictability. The Principle of Incomplete Explainability/Predictability is based upon the Principle of Less-than-Strict Determinacy.*

The **Principle of Incomplete Explainability/Predictability** states: Human comprehension is capable of understanding the world within a certain limit.

This limit is reached whenever emergence is to be understood.

Human comprehension has to

(I)   restrict deductive explanations/predictions to a limited case only; •

(II)   ascend from the abstract to the concrete by using analytical and synthetic tools in unison;

(III)   search for the adjacent necessary;

(IV)   reconstruct the concrete as a unity of essence and appearance.

Adhering to the standpoint of reflexive rationalism does not mean that deduction fails entirely in the context of explanation/prediction. It works in all cases in which emergence need not be considered. Thus, the deductive figure turns out to be a special case of understanding. Understanding seeks the adjacent necessary. In most cases, ontology does not admit more. Only in a minority of cases (no self-organisation) might the necessary condition also suffice, enabling deductive explanations/predictions.

5.2 Grasping Complexity

Moving the four research programmes from the philosophical context of the contradiction of explainability and unexplainability into the cross-disciplinary context of complex systems theory yields a sophistication with regard to another contradiction. Methods of thinking in complexity are exposed to the tension between the use of formalisation and the handling of the nonformal. Accordingly, they favour formalisation (as in the case of empiricism and cabalistics) or they favour the nonformal (as in narrativism) or they bridge the gap (as in Evolutionary Systems Methodology) (Table 5.2).

Table 5.2. Four research programmes from the system theoretical point of view.

research programmes in philosophical terms		research programmes in cross-disciplinary terms		core idea
ratio-nalism	scientism	formalism	empiricism	computation, simulation
	anthropo-socio-morphism	(quantita-tive methodo-logy)	cabalistics	
irrationalism		narrativism (qualitative methodology)		nonformal description (narratives)
reflexive rationalism		evolutionary systems methodology		gaps between phases and between levels

Formalisation is the activity of symbolising a real-world process in order to be able to operate on the symbols while considering only the rules of their combination and neglecting their referents. After having carried out the operation, you can refer to the referent again. Formal logic and mathematics are instances of such languages. They can work if and because the real-world referents are in reality linked in the strict deterministic way that formal logic and mathematics link symbols.

5.2.1 *Empiricism*

Here, "empiricism" describes the inclination to search for data that can be quantified and subjected to (computer-based) statistical tools of analysis and interpretation. Such a data processing procedure implies that the data can be formalised.

Formalisation allows applying intellectual methods provided by formal sciences such as formal logic, mathematics, or computer science; these methods involve deducing a conclusion from its premises or calculating a result or a computer operation [Krämer 1988]. Due to their nature, however, these methods of mental transformation lead unequivocally from something that is given in the mind (as a starting point) to something that follows from it in the mind (as an end point).

What works as a starting point is interpreted as a model of the cause in a real-world process; what plays the role of an end point is taken as the model of the effect. The whole transformation in the mind functions as a model of a causal relationship that is so by necessity and not contingent. These formal methods apply only in the case of a mechanical process. Mechanical processes can be mapped onto algorithmic procedures that employ clear-cut and unambiguous instructions conducted by computers as universal machines.

Like the underlying deduction procedures, mathematical operations on computers cannot reveal emergent processes in the object. Deductions, by definition, do not yield novelties. By definition, neither do algorithms or computation. The empiricist research programme becomes reductionistic if applied to emergence.

The distinction between the property "deterministic" and the property "probabilistic" concerning automata is, in this context, misleading. Probabilistic machines also rely completely on strict deterministic mechanisms in the sense defined above and are thus mechanistic. This is true despite their inclusion of, e.g., "random numbers", which are, in fact, pseudo-random numbers produced by strict deterministic mechanisms [Fuchs-Kittowski 1976, 193]. Machines do not choose. Claiming that they do would blur the distinction between the way mechanical devices work and the way systems endowed with subjectiveness (evolutionary systems, i.e., self-organising systems) act. At best we can say that probabilistic computing is a way to simulate less-than-strict deterministic processes of real-world systems; nonetheless, it is not exactly the way these processes work in the real world.

This holds true for evolutionary computing as well. Apart from using the same computer mechanisms, the computation of evolutionary processes apparently involves a mechanistic misinterpretation of Darwinian theory [see e.g. Peter Corning 2003, who is one of the critics]. This makes such computing, at best, a simulation of real-world evolutionary processes but not identical to them or an evolution itself. As Mario Bunge [2003, 152] phrased it, "things are not the same as their artificial simulates. In particular, a computer simulation of a physical, chemical, biological, or social process is not equivalent to the original process: at most, it is similar to some aspects of it." Susan Oyama [2000] compiled a thorough literature collection dealing with that problem in biology.

5.2.2 *Cabalistics*

Sometimes the application of formal methods insinuates underlying laws of nature or of society, but this is not true. In these cases, artefacts like mathematical constructs are merely projected onto reality but do not correspond to real processes and relationships. They are wrongly considered to map a real-world feature. This group of projectivistic research programmes might *pars pro toto* be called, based on their play with numbers, "cabalistics".

5.2.3 *Narrativism*

Systems modelling always confronts us with situations that are not formalisable using such an approach and hence are unpredictable. Computer science recognises the problem of the nonformal [see Fuchs-Kittowski 1992]. This explains the call for qualitative methods as opposed to quantitative ones. Qualitative methods are regarded as being the only appropriate ones for the nonformalisable.

Postmodern thinking established the narrative as paradigmatic for qualitative methodology. Accordingly, such research programmes might be termed "narrativism". Narratives are empirical descriptions that seemingly escape formalisation and make it difficult to draw conclusions.

5.2.4 *Evolutionary Systems Methodology*

What is needed is an integrative solution that accepts formalisation and qualitative handling of the informal *cum grano salis* but not their one-sided exaggerations.

Bertalanffy, in writing about isomorphies as systems laws to be found in systems of any kind [1950, 137], was aware of the fact that this also

is true for phenomena where the general principles can be described in ordinary language though they cannot be formulated in mathematical terms.

Bertalanffy shared this conviction with other representatives of General Systems Theory [see Hofkirchner et al. 2011]. Such a conviction has consequences for devising the relationship between the analytical method and its use of mathematics, on the one hand, and the exclusive use of ordinary language as a scientific method as well, on the other hand. Wherever possible, science should strive for mathematical formulations; where this is not possible, science should not belittle or dispense with natural language qualitative methods.

Taking this seriously means that nonformal, qualitative, ordinary language methods are indispensable for the scientific reconstruction of self-organising systems.

This is so because emergence in phases of evolution, including the appearance of novel qualities in developments, and differences between system levels, cannot be formalised such that a transformation leads unequivocally from one level to another. There is no simple, deductive inference that leads compellingly from premises about the system in one state (or systems) – or about one system layer – to a conclusion about the system in another state (or a metasystem) – or about another system layer. Gaps in formalisability must be accepted by the new approach to knowledge construction followed by Evolutionary Systems Methodology. It is impossible to find an operation in the mind that accomplishes the leap from one quality to another in an unambiguous and compelling manner. Thus, nonformal language must, in principle, suffice. Note, however, that this should not hamper permanent efforts in refining this language.

In phase transitions, leaps in quality exist between the state of the system at one point of time t and the following state at time t+1. Such leaps also exist between systems in the state before a metasystem transition and the metasystem state itself. This condition has to be covered by a diachronic saltation in method (Figure 5.6).

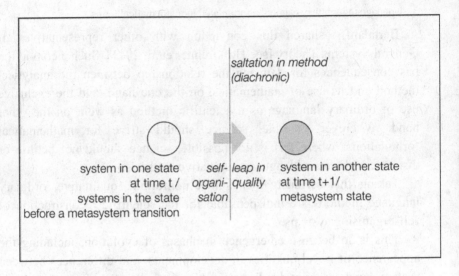

saltation in method
(diachronic)

system in one state   self- |leap in   system in another state
at time t /   organi-|quality   at time t+1/
systems in the state   sation   metasystem state
before a metasystem transition

Figure 5.6. Diachronic method saltation following the leap of a phase transition.

Level shifts involve leaps in quality between one layer n of the system and the adjacent layer up or down n+1 or n-1. Such leaps must be covered by a synchronic saltation in method (Figure 5.7).

Herbert Simon wrote, in his *Society for General Systems Research Yearbook* 10 contribution on *The architecture of complexity*, about the problem of how to link system layers like the macro- (whole) and the micro-level (parts) [François 2004, 103]:

> In such systems, the whole is more than the sum of the parts, not in an ultimate, metaphysical sense, but in the important pragmatic sense that, given the properties of the parts and the laws of their interaction, it is not a trivial matter to infer the properties of the whole. In the face of complexity, an in-principle reductionist may be at the same time a pragmatic holist.

Bertalanffy pointed to the benefits, if not necessity, of change in the method when jumping from a micro- to a macro-level [1950, 141]:

> If you cannot run after each molecule and describe the state of a gas in a Laplacean formula, take, with Boltzmann, a statistical law describing the average result of the behaviour of a great many individual molecules.

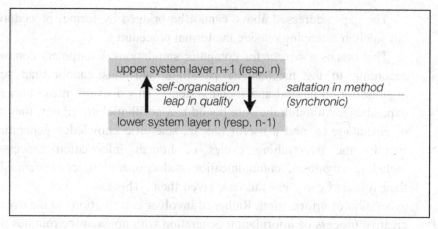

Figure 5.7. Synchronic method saltation following the leap of a level shift.

This calls for defining the type of comprehension that evolutionary systems need:

Evolutionary Systems Methodology. *Evolutionary Systems Methodology is that research programme (programme of comprehension) that applies the Saltation Principle. The Saltation Principle is based upon the Principle of Incomplete Explainability/ Predictability as well as the Principle of Self-Organisation.*

The **Saltation Principle** states: The research programme is to use methods for comprehending self-organising systems that acknowledge the unpredictability (and the accompanying lack of retrodictability) of evolutionary and developmental trajectories as well as nested hierarchies. Nonetheless, that programme is capable of comprehending qualitative leaps occurring (I) during phase transitions as well as (II) during level shifts.

In particular, the Saltation Principle consists of two principles:

(I) The **Principle of Diachronic Leaps in Quality** states with regard to diachrony: Include saltations when going from one state of a system to the subsequent state or from protosystems to the metasystem!

(II) The **Principle of Synchronic Leaps in Quality** states with regard to synchrony: Include saltations when going from one system layer to the next upper layer (and thus *vice versa*)!

The gaps addressed above cannot be bridged by formal procedures, but solely by stepping outside the formal procedures.

This means a *caveat* for computer simulations. Computers compute according to the mechanistic paradigm and thus cannot map self-organisation in natural and social systems. This does not mean they are expendable. Although they are bound to algorithmic procedures, they are of advantage to, and a useful link in, scientific knowledge generation. Inmidst the overarching cycles of human information processes, including cognition, communication and cooperation, computers play their role and carry out the task given them. This task is not *per se* the generation of information. Rather, it involves contributions to the overall creative process of information generation with noncreative routines that gain their full potential only in the greater context. This situation can be compared to the field of logic. We clearly make use of deductive logic but are aware of the power of the unformalisable.

Chapter 6

Objects are Subjects are Objects...
Reflections in a Creative Universe

The environment contains no information; the environment is as it is.

– Heinz von Foerster: Understanding Understanding, 2003 (1972) –

In fact, what we mean by information – the elementary unit of information – is a *difference which makes a difference* [...]

– Gregory Bateson: Steps to an Ecology of Mind, 1973 –

How is it possible for a physical thing – a person, an animal, a robot – to extract knowledge of the world from perception and then exploit that knowledge in the guidance of successful action?

– Daniel Dennett: When philosophers encounter artificial intelligence, 1988 –

Behavior has no opposite; one cannot *not* behave [...] if it is accepted that all behavior in an interactional situation has message value, i.e., is communication, it follows that no matter how one may try, one cannot *not* communicate.

– Paul Watzlawick, Janet Beavin Bavelas, and Donald D. Jackson: Pragmatics of Human Communication, 1967 –

From the beginning, we've been yanked together by the tug of sociality. Three and a half billion years ago, our earliest cellular ancestors, bacteria, evolved in colonies. Each bacterium couldn't live without the comfort of rubbing against its neighbors. [...] From the beginning, we living beings have been modules of something current evolutionary theory fails to see, a collective thinking and invention machine.

– Howard Bloom: Global Brain, 2000 –

Having outlined several interdependent principles of a new philosophy (POE, including ways of thinking) and a new cross-disciplinary paradigm (EST), the foundations of the UTI framework are laid. The next step is to erect the scaffolding. The discussion begins with philosophical implications and continues with the cross-disciplinary, general implications of the foundations for the scaffold. In parallel, the scaffold is built up.

6.1 Dissolving Capurro's Trilemma

As a start, the review of the classifications of information concepts/theories seems to support the assumption of a multitude of approaches that are diverse and irreconcilable, with no possibility of consolidation. The attempt to construct order from noise to turn the babel into polyphony faces a situation known as "Capurro's Trilemma"[a].

Capurro's Trilemma posits [Capurro et al. 1997]: in attempting to define and determine what "information" means throughout the disciplines (as well as in everyday thinking) and what it should or could mean, one faces a logical situation that offers three options, none of which are satisfactory (Table 6.1).

(1) The first option is: there is only one meaning of the term "information"; it means the same thing regardless of the field of application. This option is called "synonymity", because the terms are synonyms (also called "univocity").

(2) The second option is: there are several meanings of the term "information"; they are similar to a particular meaning, which serves

---

[a] In December 1990 I by chance met the information philosopher Rafael Capurro during a gathering of the *Arbeitskreis Informatik und Philosophie* of the German *Gesellschaft für Informatik* in Bremen, where – on behalf of Peter Fleissner who later became head of the working group I participated in – I presented a draft research proposal on information concepts. At this meeting, Capurro made an intervention concerning the trilemmatic situation. In the following years, after establishing a closer relationship to Capurro, Fleissner and I took to calling it "Capurro's Trilemma". Capurro, however, insists on merely having interpreted Aristotle and applied his thinking to information concepts.

as a standard of comparison. This option is called "analogy", analogical reasoning, because the terms are analogies.

(3) The third option is: there are several meanings of the term "information", all of which differ from each other. This option is called "equivocity" because the terms are equivocations.

Table 6.1. Capurro's Trilemma.

	information terms have...
synonymity/ univocity	... exactly the same meaning
analogy	... similar meanings
equivocity	... completely different meanings

In fact, no option meets the demands for scientificity. Synonymity does not meet them because information in one domain would not differ from information in a different domain – a premise which has long been contested. Analogical reasoning does not meet them either because there is no agreement on the *primum analogatum*, the standard of comparison. Nor does equivocity meet them, because the babel of languages which are not communicable would mean the end of scientific enterprise as such.

Does this mean that we are stuck and that there is no solution to the trilemma?

Taking a closer look reveals the connection between the options given and the scientific strands identified in earlier chapters:

(1) synonymity is the option preferred by a "hard" science perspective;

(2) analogy goes with the "soft" science perspective; and

(3) equivocity firmly encases each of those perspectives as a separate pillar (two cultures).

The solution to the trilemmatic situation lies in recalling that each of the three perspectives, the "hard" science perspective, "soft" science perspective and two-cultures perspective, reflects one distinctive way of thinking, that is,

(1) the "hard" science perspective reflects reductionism,
(2) the "soft" science perspective reflects projectivism,
(3) and the two-cultures perspective reflects disjunctivism.

We need only to introduce the fourth way of thinking – integrativism (Table 6.2).

Table 6.2. Four conceptualisations of information.

	information praxiology	information ontology	information epistemology
reductionism (synonymism)	objectivism	materialism	externalism
projectivism (analogism)			
disjunctivism (equivocalism)	subjectivism	idealism	internalism
integrativism (historical and logical conceptualism)	subject-object dialectics	emergentist materialism	perspective-shifting methodology

Reductionism reduces the meaning of information to one and the same meaning; projectivism projects a particular meaning of information to all the other meanings; disjunctivism disjoins every meaning from any other meaning of information.

6.1.1 *Synonymism*

The reductionist way of thinking about information approaches information as a "hard" fact, as a given to whatever science (this holds for everyday thinking too), as one single thing.

Objectivism. Praxiologically, information is treated as the object of any type of practical action by humans, i.e., in the context of social (cultural, political, economic) steering and intervention, in environmental management, in the usage and the design of technology. It is a something that can be handled, in particular stored, retrieved, distributed, transmitted, received, and processed.

Materialism. Ontologically, this something is conceived of as an object of the real-world, be it in the realm of human society or in the wider physical world. It need not necessarily be a substance in its own right (like matter is said to be) in order to qualify for a materialistic ontology. It is sufficient to consider it as a derivative of matter, that is, as a property of matter (such as structure), which holds for signal transmission as well. Either way, it is considered a material object.

It is worth noting that the nature of information in this context is perceived as rather fixed and static than fluid or dynamic.

Externalism. Epistemologically, this material object is the object of inquiry by empirical and formal-scientific methods, all of which are carried out in the so-called third-person perspective, i.e, from the point of view of an outside observer. Formalisms are developed to measure the structure of matter or signals.

6.1.2 *Analogism*

The projectivistic way of interpreting information is to approach it as a "soft" phenomenon.

Accordingly, in analogy to subjective human action that is carried out by human actors and considered to be an information process, subjective activity of the same kind is postulated to be carried out by any agents that populate the world.

Subjectivism. This is, praxiologically, a projection of human subjectivity in information processes onto processes with nonhuman entities whose degree of subjectiveness might be questioned.

Idealism. This is, ontologically, a projection of a presumed intentional nature of human information processes onto the nature of processes caused by nonhuman entities whose intendedness is questionable; this explains the fluidity and dynamicity of all information in an idealistic view.

Internalism. This is, epistemologically, a projection of subjective methods of gaining insight into the phenomenon of information in humans onto the study of nonhuman information processes.

6.1.3 *Equivocalism*

The disjunctivist interpretation of information reserves a distinct category for the subjective (human) action of information: praxiologically, in the context of a social discourse involving humans only; ontologically, as a human phenomenon only; and, epistemologically, as subject to subjective interpretation only.

6.1.4 *Historical and logical conceptualism*

The above considerations pave the way for circumventing the trilemma. Introducing the fourth way of thinking – integrativism – immediately dissolves it.

In the objective view, reductionism treats information as an object with an exclusive focus on the structural aspect, and in the subjective view, projectivism and disjunctivism regard information as an agent's subjective action or activity with an exclusive focus on the process aspect. Integrativism reveals information as a subjective process that is constrained and enabled by information as an objective structure (that is produced by information as a subjective process...). Processes do "process" structures, and structures do structure processes. This prompts the processual and the structural views to fuse to the extent that the subjectivistic and objectivistic, idealistic and materalistic, internalistic and externalistic views are mediated and the two cultures of the "soft" and the "hard" sciences are prepared for bridging.

We are now in the position to recognise the shortcomings of reduction in the synonymity case, to see the one-sidedness in the projection done by analogy, and to set aside the disjunction that denies common ground in the case of equivocity. At the same time, we can do justice to what they posit reasonably.

According to strict synonymism, we need to accept that information is something objective (not belonging to a subject) that can be measured independently. And according to strict analogism and equivocalism, information is something subjective, that is, inextricably linked to a subject that is human. A UTI cannot be satisfied by such one-sided views. An integrative science of information has to consider both the objective, material, external, and subjective, ideal, internal aspects of information and at the same time overcome synonymism and analogism and equivocalism.

The objectivistic, materialistic and externalistic outlook is correct in stating that information is an objective matter of fact in the real world and not merely human imagination. The subjectivistic, idealistic and internalistic outlook is correct insofar as it states that information occurs only if there is freedom of choice for that to which generating and utilising of information is attributed, which means it is a subject. Regarding the first outlook, however, we have to limit the scope of objects within which information is said to be found to those objects exclusively that take the role of subjects. Regarding the second outlook, we have to enlarge the scope of subjects from that of humans exclusively and include non-human ones, too, as quasi- and protosubjects.

What a UTI seeks is a concept that is simultaneously as abstract as necessary but as concrete as possible. The more abstract a concept, the poorer it is by intension and the larger by extension. The more concrete a concept, the richer its intension and the smaller its extension.

On the one hand, the concept theorises what all information processes have in common, but should not reduce them to an abstract formalism that can subsume every case under a meaningless definition. On the other hand, it should cover each empirically detectable individual information process but not hypostatise its unique particularities into a concretistic notion.

That is the real challenge.

The desired concept is a historical and a logical concept in one. It has to be historical in that it explains the historical appearance of different information processes and different information structures. It has to be logical in that it shows how different manifestations of information processes and structures are logically linked (which does not mean a deductive scheme).

We need a concept that is concrete in the sense of what can be reached when ascending from the abstract. This calls for taking advantage of the ascendence from the universal to the particular as pointed out in the Principle of Incomplete Explainability/Predictability, which is based upon the Principle of Less-than-Strict Determinacy, which in turn builds upon the Principle of Limited Controllability – all of which depend on the Principle of Unity-Through-Diversity.

Such a conceptualisation would be flexible enough to balance the universal and the particular, to do justice to both, to relate them so as to render the universal in need of as well as capable of being completed by the particular and, in turn, embed the particular in the universal [see Hofkirchner 2004].

Reductionistic unification would reduce the particular to the universal by stating "The Particular is (nothing but) Universal" and by assuming that the universal is the necessary as well as sufficient condition for the particular. This is true of all kinds of subsumption. They overlook what goes beyond that which subsumes. Unification by projection would project the particular onto the universal and postulate "The Universal is (nothing but) Particular", meaning that the particular is not only necessary but also sufficient to yield the universal. This holds for those illusions that extend commonalities to a realm where they do not exist. The disjunctive way of thinking would dissociate the particular from the universal by presuming "The Particular and the Universal are Disjoint" and would, in doing so, insinuate that both notions contradict each other. This causes the particular to fall apart because there is no unifying bond. Each of these three ways of thinking is one-sided because, by relying on the formal-logical figure of falsely stated necessary and sufficient conditions or of contradiction, they all focus on the mutual dependence of the sides or on the condition of being opposites. None comprise the full range of what is characteristic of any dialectical relation.

It is only the fourth way of thinking that integrates as well as differentiates the particular and universal. This point of view may be formulated as "The Particular Sublates the Universal" – "sublation" in the threefold Hegelian sense denoting suspending, saving and elevating altogether:

(1) the particular suspends the universal; being the opposite of the universal, the particular contradicts the universal and transcends it;

(2) the particular saves the universal; the particular depends on the universal, the latter being the necessary, but not sufficient condition for the particular; the particular is based upon the universal;

(3) the particular elevates the universal to another level; in an asymmetrical effort, the particular turns the universal, as a consequence, from an abstract universal into a concrete-universal.

The concrete-universal is the unity that overarches the diversity of the particulars. Aristotle paved the way for a dialectics of the universal and the particular by establishing specification hierarchies via *genus proximum* and *differentia specifica*. The whole tree can be considered to represent the concrete-universal, and each ramification to specify one particular instantiation of the universal by making the abstract concrete.

Hierarchies that specify the being are the logical way of grasping the history and genesis of becoming (unity of being and becoming). According to that conceptualisation approach, which is logical and historical at the same time, a UTI seeks a concrete-universal concept of information rather than an abstract one.

Subject-object dialectics. Objects and subjects are defined by (1) mutual exclusion as opposites (2) depending on each other (3) in an asymmetrical relation.

(1) Objects are subject to subjects, while subjects subject objects.

(2) Objects do not exist unless subjects exist, and *vice versa*.

(3) Objects and subjects are bound together by the process in which subjects interact with objects.

Such a relationship is termed dialectic.

The subject-object dialectics can be visualised by a cycle in which the subject couples to the object (Figure 6.1).

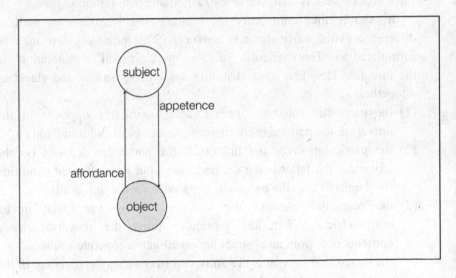

Figure 6.1. Subject-object dialectics.

On the one hand, the object bears significance for the subject in so far as its objective properties suit subjective functions – this is called "affordance" elsewhere [Gibson 1950, 1966, 1979]. On the other hand, the subject designates the object for serving it one way or another because it needs to reach out for that which it makes into an object and it needs to approach the object in a subjective way. In a different context, this has been termed "appetence".

This very relationship harbours the origins of information: because subjective appetence relates to objective affordance, we have subjective signification designating objective significance – "signification", "designation" and "significance" describing a field of latency for information to emerge. It is this relationship by which a subject relates itself to an object via its own activity, from which information emerges (Figure 6.2).

Humans are subjects. Through interference with their human and non-human surroundings, they produce objects. These objects tend to resist ("object" to) becoming subject to humans because they are characterised by inertia. Praxis is the ongoing process of subjecting objects to humans while factoring in inertia. More often than not, however, this inertia turns out to reflect the inert nature of the objects but the level of subjectiveness they inhere according to their nature as a type of subject. In that respect, humans must come to acknowledge that they interfere with quasi- and proto-subjects and that it is up to us, as humans and fully-fledged subjects, to minimise the frictions occurring in the course of interaction between different kinds of subjects. This would allow us to better pursue our own interests.

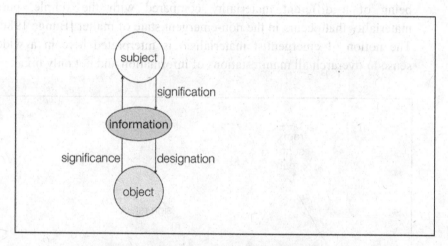

Figure 6.2. Subject-object-dialectics in information.

Emergentist materialism. According to how POE is designed, the materialism–idealism divide can be derived from the object–subject divide in that matter is objective and ideas are subjective. This helps to clarify that matter and information, like ideas, belong together as objects and subjects.

An answer that goes beyond (materialistic and idealistic) monism and (if you like, interactive) dualism is the answer of dialectics. Dialectics concurrently recognises identity and difference of matter and information: it recognises identity, given the difference, because this identity enables these different sides to interact; and it recognises the difference, given identity, because this makes it possible to differentiate matter and information as different specifications of an identical, common genus.

This common ground must be one of the two sides: otherwise the intention of integrativism would be flawed. And it should be that side that shows evidence of giving rise to the other, though in a non-derivative way.

Emergent(ist) materialism is known as a philosophy of mind that fulfils the criterion. Matter is the common ground but leaves room for emergent properties and events. This is because mind is regarded as being of a different materiality compared with the simple, pure, materiality that occurs in the non-emergent state of matter [Bunge 1980]. The notion of emergentist materialism is interpreted here in a wider sense to overarch all manifestations of information and not only mind.

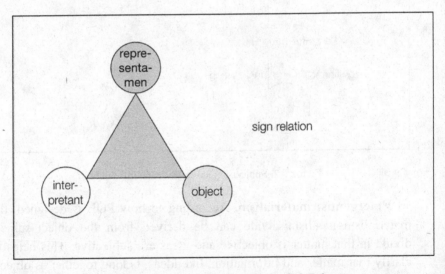

Figure 6.3. The semiotic triangle after C.S. Peirce.

In the present context it makes no difference whether we talk about information processes or the results of information processes, i.e. structures. The process of information generation generates information, that is, it produces, and ends up in, an information structure that remains genetically linked to its production process. That structure, in turn, functions as a frame for another process when utilised.

Accepting the above, it is useful to turn to semiotic arguments and learn how they conceive information as process and structure. Semiotics deals with semiosis, which is sign processes that make use of signs as structures. Signs are fundamentally defined as relationships.

What is a sign? In everyday life the following definition is widely used: A sign is something that stands for something else. It can be argued that this common-sense definition is appropriate and capable of development in scientific respect, too. In science, traditionally, a sign is defined, on the one hand, by dyadic semiotic schools either as a unity of carrier and meaning or as a unity of carrier and object and, on the other hand, by triadic schools as a unity of carrier, meaning and object altogether, whatever the terms for the correlatives may be [Nöth 2000, 136-141].

As an example, triadic semiotics in the tradition of Charles Sanders Peirce [1983 and 2000] recognises the "representamen" (the sign in a narrow sense as some kind of carrier), the "interpretant" (which means the "meaning" of the representamen and is not to be confused with an interpreter), and the "object", which altogether form the so-called semiotic triangle (Figure 6.3).

In both cases, however, the notion of sign is implicitly linked to the notion of subject. Thus, speaking of a carrier that conveys meaning or refers to an object requires speaking of a subject, too. Only if a subject is assumed does it make sense to assume something as a carrier of meaning, because that something is not a carrier unless it serves a subject. Moreover, only if a subject is assumed does it make sense to assume something like meaning. This is because meaning is always meaning to some subject. Last, but not least, only if a subject is assumed does it make sense to assume an object, because things are not objects until they are subject to a subject.

Therefore it makes sense to revisit traditional semiotic schools, including the so-called semiotic triangle, from the viewpoint of subject-object dialectics (Figure 6.4).

It is helpful to recall the subject-object dialectic cycle and take into consideration that a subject never relates directly to an object. Its relation to the object is always mediated. It construes the means of mediation. In the course of the subject's acting upon the object, the subject gives rise to something new by which it mediates itself with the object – the sign. The sign is a means for the subject to accord its appetence for the object with the affordance of the object. The appearance of the sign (*signans*) turns the subject into a signmaker (*signator*); the appetence relation into a signification process (*significatio*) that produces the sign and a designation process (*designatio*) that refers to the object; the object into a something (to be) signified (*signandum/signatum*); and the affordance relation into one that conveys the significance the object bears for the subject (*significantia*).

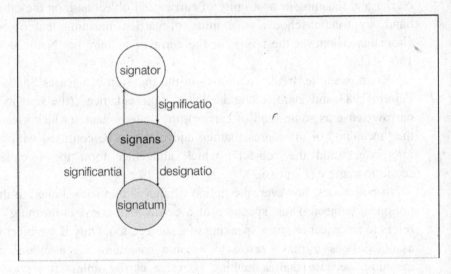

Figure 6.4. A semiotic elaboration of the subject-object-dialectics.

This gives rise to a different semiotic triangle if we give the subject-object dialectic figure a triangular shape (Figure 6.5): the subject (*signator*) generates and utilises the sign. The object is that part of the relation for which a sign is (to be) produced (*signandum/signatum*). And the sign is a product of the subject's own activity that is to stand for the object (*signans*). The engagement of the sign actualises the potential significance of the object to the subject (*significancia*) by assigning a sign to the object (*significatio* and *designatio*). Thus it lets meaning emerge.

The triangle is made up of the three correlatives signmaker (*signator*), (to be) signified (*signandum/signatum*) and sign (*signans*). Note that there is no relation between *signandum/signatum* and *signans*, unless there is a *signator* that establishes the relation. Even more important: after the sign relation has been established, the *signator's* interchange with the *signatum* is *signans*-mediated and henceforth channeled by the *signans*. The *signans*, once emerged, is the new quality that exerts dominance over the *signator's* relation to the *signatum*.

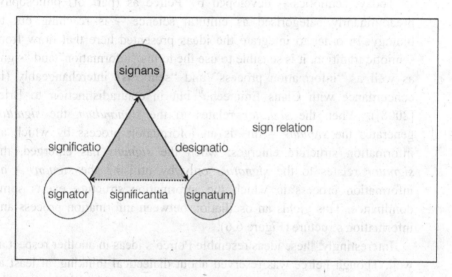

Figure 6.5. A semiotic triangle compatible with subject-object-dialectics.

The triangle sketched here differs from the traditional semiotic triangle in many respects. The nodes are different, and so are the links. In the traditional semiotic triangle there is no subject, but meaning instead. In the triangle given here, the subject (*signator*) is an indispensable part of the relationship, and meaning is the relation from the subject to the object when mediated through the sign (*significatio, designatio*). In the traditional semiotic triangle there is a carrier for the sign and the sign is a thing made up of all of the three nodes of sign carrier, meaning and object of reference. The present triangle does not differentiate between a sign and its carrier. The sign is an intermediate entity at the subject-object interface that is constructed in the conflict between the subject and the object. Only in the course of evolution is there a reification of the sign into a carrier. The traditional semiotic triangle can thus be considered to describe a special case in the variety of semiosic processes. It may be considered embedded in the triangle presented here, that is, as a node provided with the properties of referring to the subject (meaning) and the object (representation) and with the potentiality of existence separate from the subject.

Today, semiotics – developed by Peirce as (part of) philosophy, predominantly categorised as cultural science – is reaching out to biology. In order to integrate the ideas presented here that draw from semiotic tradition, it is sensible to use the terms "information" and "sign" as well as "information process" and "semiosis" interchangeably (in concordance with Claus Emmeche[b] but in contradistinction to Brier [2008]). When the *signator* relates to the *signandum*, the *signator* generates the *signans* – this is an information process by which an information structure emerges; when the *signans* has emerged, the *signator* relates to the *signatum* only by utilising the *signans* – an information process in which the information structure exerts some dominance. This yields an oscillation between information process and information structure (Figure 6.6).

Interestingly, these ideas resemble Peirce's ideas in another respect as well. Though Peirce was reserved about dialectical thinking, at least as far as thoughts of Hegelian origin are concerned, he himself developed a classification of relations according to the categories of firstness, secondness and thirdness, which express dialectical thinking and which he defended in a dispute with Bertrand Russell. Russell, from his rather formal point of view, did not see the point Peirce was making when contending that every relationship – those of a pretended "fourthness", "fifthness" or "n-ness" – can, finally, be dissolved into relations along the primary categories. Firstness represents an individual relating to nothing than itself, secondness two individuals relating to each other, and thirdness a tripartite relationship in which a third establishes the relation between the other two. It is with this thirdness that novelty emerges – in

---

[b] Personal communication 2005.

the context of a UTI this can be interpreted as the structural moment of information processes.

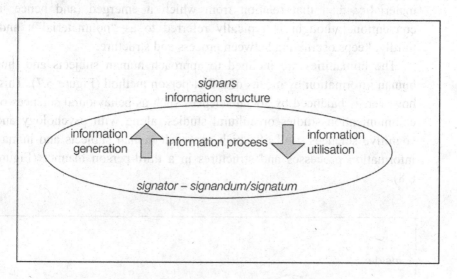

Figure 6.6. Information structuration process by incorporating semiotic terms.

In summary, information is a relation, not a material entity or an ideational event.

It is a mediated relation, not a direct relation between something (a subject) and another thing (an object). It is mediated by a third thing, which makes for the unique nature of the whole relationship.

Perspective-shifting methodology. Clearly, in the epistemic context of information, repeated shifts in taking perspectives[c] are ineluctable. This is because information, regarding praxic context, emerges in the field of subject-object dialectics, in which the subject's material appetence relation to the object provides the possibility of being converted into a signification relation and in which the object's material affordance relation to the subject provides the possibility of being converted into a significance relation. This is also because information, regarding ontic context, is a material emergent that, firstly, realises these

---

[c] Ludwig von Bertalanffy called the epistemological stance of his General System Theory "perspectivism" [1965], which I draw upon.

conversion possibilities through designation (mediation via a sign structure); secondly, represents a relation that is of a higher-order materiality than that relation from which it emerged (and hence in conventional thought is typically referred to as "nonmaterial"); and, thirdly, keeps oscillating between process and structure.

The humanities are inclined to approach human subjects and thus human information by means of a first-person method (Figure 6.7). This, however, is balanced by social sciences such as behavioural sciences or communication studies or cultural studies, along with psychology and cognitive sciences – all of which approach human subjects and human information processes and structures in a third-person manner (Figure 6.8).

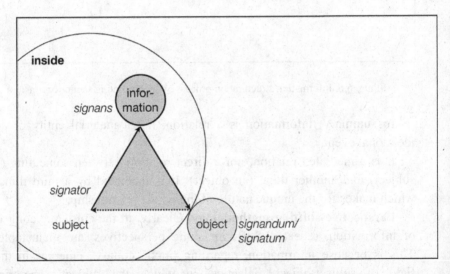

Figure 6.7. Internal view of information using semiotic terms.

Human subjects seem to be researchable from within and from without. These are complemented by nonhuman subjects, quasi- and protosubjects. The latter seem to be open to inquiry only from without. If, however, we suggest a genealogical lineage from proto- to quasi- to fully-fledged subjects, and thus a deep connection between all kinds of subjects, why should an investigation of the predecessors of human subjects from within be impossible? We merely need to accept that empathy comes stepwise and that *homo sapiens*, who features the most sophisticated empathic capabilities, can try to apply these capabilities in grades. We can conclude: subjects and their information processes and structures might be studied from both an external and internal perspective. Indeed, in order to grasp the whole phenomenon and not omit any essential feature, they must be studied from both perspectives. Objects are in fact subjects that can be studied as objects and as subjects as well. Objects and subjects are two sides of the coin. Switching back and forth between the inside stance and the outside stance reveals emergent qualities of the information phenomenon.

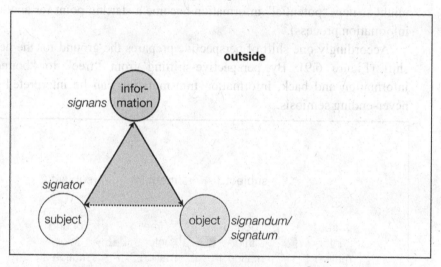

Figure 6.8. External view of information using semiotic terms.

Processes and structures of information transmute into one another. In order to capture the different mutations of information, shifts in the perspective taken are appropriate here too. One shift is from viewing processual, fluid, "free" information (as a process on behalf of a subject) to viewing structural, frozen, "bound" information (as that structure that is produced by this very subject). This would help understand the structuration (concretion) of information.

Since this information structure represents a potential from which another information process can depart, another shift is from grasping the freeze to grasping the leaking of information after the defrosting, melting, liquifying. This would help understand the processualisation (diffusion) of information, the reaching out of the "potential", "bound" information merely by virtue of the subject's selecting it for showing up as "free" information in some medium.

A next shift would help understand the "actualisation" of the "potential" information by another subject – a step in which the process is frozen down into "bound" information again, but finds itself in another structure, in a new structure generated by the new subject (which, in turn, might as new "potential" information become a starting point for another information process).

Accordingly one shift of perspective prepares the ground for the next shift (Figure 6.9). By perspective-shifting from "free" to "bound" information and back, information transmutation can be interpreted as never-ending semiosis.

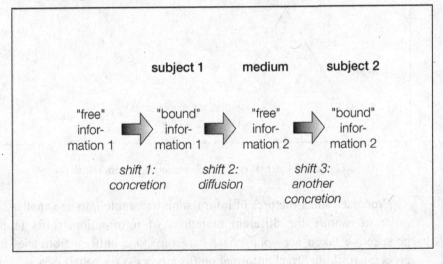

Figure 6.9. Views of information as "free" and as "bound".

This leads to the following definitions:

Historical and logical conceptualism. *Historical and logical conceptualism is that way of thinking about information that applies the Principle of the Historical and Logical Account of Information. The Principle of the Historical and Logical Account of Information is based upon the Principle of Incomplete Explainability.*

The **Principle of the Historical and Logical Account of Information** states: The meaning of the concept of information has to comprehend both what different manifestations of the phenomenon of information have in common and what is unique to them. Historical manifestations of information descend from earlier manifestations but do not derive from them logically. With each historical manifestation that is to be conceived of, the concept of information is enriched by features not characteristic of it so far and extended so as to make the universal and the concrete unify in order to include the manifestation in the extension of the meaning. No concrete concept of information can be deduced from a more abstract concept, but an abstract concept can be deduced from a more concrete one.

The information concept desired in a UTI is a concrete universal.[d]

---

[d] Examples for such an information concept can be found not only in the biological classification of species but also in social sciences. An example of political economy is the development of capitalism. It's clear that this economic formation underwent several transformations (sometimes regarded as mutations into a different economic system that is not capitalist any more, but the reflection of the financial crisis that caused the current economic crises brought those speculations back down to earth). E.g., the following events have been argued in favour of transformations within capitalism: the development from free competition towards monopolies, the development of a close relationship between nation states and national monopolies, the development of transnational corporations, the development of the preponderance of financial capital over industrial capital in the course of globalisation and informatisation. Each development was, in a way, unpredicted and deemed to modify the "essence" of capitalist principles but not replace it fully. It might characterise a new stage in the evolution of capitalism, as the latest notions of "global capitalism" and "informational capitalism" insinuate.

6.2 Emergent Information

The philosophy of information considerations discussed above are useful to approach an answer to the question of the place that information has in the universe and the role that creativity plays (esoteric answers range from an early book of Young [1974] to a more recent book of Görnitz and Görnitz [2002]).

According to a quote from Bateson, which advanced to his famous definition of information (see chapter 1), information is "a difference which makes a difference" [1972, p. 453]. In the framework of what has been outlined above about evolutionary systems and about the objective and the subjective, this saying might be reformulated as follows: we can speak of information if there is a difference in the environment of a self-organising system (the objective aspect) that makes a difference to this very system (the subjective aspect); a difference in the environment might be instantiated by an event or an entity, and the difference made to the system might manifest itself as a change in its structure, state or behaviour, which might be observable.

Using the notion of "variety", which plays an important role in W. Ross Ashby's cybernetic theory[e], the Russian philosopher of information A. D. Ursul had highlighted the intrinsic connection between information and difference in a similar manner. He defined information as "reflected variety" [1970, 166, 214 – translation W.H.]; information depends on variety and reflection: it is "variety that one object contains from another object" [1970, 166 – translation W.H.], "variety that is contained in an object in relation to another object (as result of their interaction)" [1970, 214 – translation W.H.].

Note that self-organisation itself is also due to objective and subjective factors, as underlined in the following definition [Halley and Winkler 2008, 12]:

---

[e] The Law of Requisite Variety states that a system is dynamically stable if its variety (number of states), i.e. the variety of its control mechanism, is greater than or equal to the variety of (the input from) another system, i.e. the variety of a system to be controlled.

Self-organization is a dissipative nonequilibrium order at macroscopic levels, because of collective, nonlinear interactions between multiple microscopic components. This order is induced by interplay between intrinsic and extrinsic factors, and decays upon removal of the energy source. In this context, microscopic and macroscopic are relative.

Accordingly, the very process of self-organisation fulfils the interpretation of Bateson's definition as well as the definition of information by Ursul. This is because self-organisation refers to an event or an entity in the environment of the system which represents a difference; it is a creative activity of the system in the course of which novelty is produced in its structure, state or behaviour that is related to the difference and marks a difference in the development of the system. In that vein, self-organising systems display information generativity. Information is produced in each self-organisation process.

Furthermore, self-organisation is a negentropic process because order is produced by it and the production of order is, by definition, a negentropic process. What then makes a difference for a system is whether or not a difference, or variety, can be functionalised by the system for its negentropic process of building up order. Thus, information is intrinsically connected to negentropy and organisation, as pointed out by Morin [1992, 350]:

Information is what allows negentropy to regenerate organization which allows information to regenerate negentropy.

Or [368]:

Information is what, starting from an engram or sign, allows negentropy to generate or regenerate negentropy on contact, in the framework or at the heart of an ad hoc negentropic organization.

Information is functional for the system's organisation.

There is another feature that neatly fits in the overall picture. Semiotics stresses the arbitrariness of signs produced. Because an object is something that is subject to mere determination by something else, and because a subject is something that "objects" to mere determination by something else, the generation of information is tantamount with drawing a self-made distinction by the irreproducible, irreversible, irreducible, unpredictable build-up of order during the process of self-organisation.

The paradigm shift from the mechanistic worldview cognisant of objects only towards a more inclusive view of a less-than-strict, emergent, and even creative universe inhabited by subjects too, gives us the tools to connect the notion of information to the idea of self-organisation; it is the very idea of systems intervening between input/cause and output/effect and thus breaking up the direct cause-effect-relationships of the mechanistic worldview that facilitates, if not demands, the notion of information. This is because information is bound to the precondition of subjects and their subjective agency. Self-organising systems that transform the input into an output in a non-mechanical way, that is, in the context of an amount of degrees of freedom undeniably greater than that of a one-option only, are subjects. And each activity in such a context, each acting *vis-à-vis* undeniable degrees of freedom, equates with the generation of information: the act to discriminate, to distinguish, to differentiate, is information.

Self-organisation therefore stands at the beginning of all information insofar as the system 1) selects one of a number of possible responses to a causal event in its environment, 2) shows preference for the particular option it chooses to realise over a number of other options, 3) decides to discriminate.

We can conclude: information is involved in self-organisation. Every system acts and reacts in a network of systems, elements and networks, and is exposed to influences mediated by matter and/or energy relations. If the effects on the system are fully derivable from, and fully reducible to, the causes outside the system, no informational aspects can be separated from matter/energy cause-effect relations. However, as soon as the effects become dependent on the system as well (because the system itself contributes to them), as soon as the influences play the role of mere triggers for effects becoming self-organised by the system, as soon as degrees of freedom intervene and the reaction of the system is unequal to the action it undergoes, the system produces information (see Haken [1988]). Information is created if the number of effects exceeds that of causes in a system. Information occurs during the process in which the system exhibits changes in its structure, state, or behaviour [Fenzl and Hofkirchner 1997], i.e., changes brought about by the system. Information is created by a system if it is organising itself at any level.

The term "reflection" is reintroduced here to distinguish this kind of self-organised, informational reaction (emergent) from a reaction of the stimulus-response type (mechanical). The term differs from that in the sense of a naïve realism. "Reflection" here does not comprise mechanical mirroring but deliberation on the human level along with all informational processes and their results on nonmechanical prehuman levels. This is in line with the German term "Widerspiegelung", which in the Hegel-Marx tradition was a dialectic one and, as the philosophical writings of Vladimir I. Lenin tried to insinuate, could and should be considered a fundamental property of all matter [1977, 53]. We live in a reflective universe – one made up of reflective systems that increasingly reflect the universe (hence the idea that the universe, in the guise of human systems, comes to reflect itself).

In a figurative sense, information can be viewed as the result of this process, as that which is new in the structure, state, or behaviour. Insofar as this new feature in system A may serve to stimulate self-organising (and therefore informational) processes to produce new features in system B, we can speak of information in a metaphoric sense – as if it were something to be sent from one system to another.

In summary, information is a valid concept in situations where 1) the deterministic connection between cause and effect is severed, 2) a system's own activity comes into play, and the cause becomes the mere trigger of self-determined processes in the system, which finally lead to the effect, and 3) the system makes a decision and a possibility is realised by an irreducible choice.

Since information generation is a process that allows novelty to emerge, it goes beyond a mechanical process that can be formalised, expressed by a mathematical function, or carried out by a computer.

We can define information in terms of evolutionary systems theory, supported by the semiotic terminology developed above:

Information. *Information I =def. relation such that*

(1) the order O built up spontaneously (signans; the sign)

(2) reflects some perturbation P (signandum/signatum; (to-be-)signified)

(3) in the negentropic perspective of an Evolutionary System s_e (signator; the signmaker).

"Perturbation" is the term Maturana and Varela use to depict the influence from outside on self-organisation[f].

Information is generated if self-organising systems relate to some external perturbation by the spontaneous build-up of order they execute when exposed to this perturbation. In the terms of triadic semiotics, the self-organising systems, by doing so, assign a signification to the order and make it a sign which stands for the so signified perturbation (Figure 6.10).

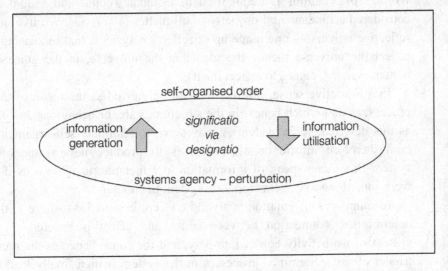

Figure 6.10. Information structuration process in EST and semiotic terms.

Figure 6.10 shows the interplay between the agency of a systems agent (*signator*) and the perturbation it suffers (*significandum/ significatum*) as processes mediated via the self-organised order as structure (*signans*). The signification-significance relation between the system and the perturbation is duplicated, becomes independent, takes on a life of its own, when becoming reified in the sign and thus upgraded to a tripartite relationship.

---

[f] I use the term despite the radical-constructivist interpretation that belittles that influence as a mere modification of self-organisation processes that are deemed to be essentially autonomous – an interpretation that tends to cut the system free from its environment.

Thus, the process of self-organisation coincides with the process of information-generation (sign-production). The respective results also coincide. The concepts of self-organisation and information turn out to be co-extensive.

6.2.1 *Evolutionary types of reflection: the Multi-Stage Model*

EST (and a UTI as an extension thereof) can play the role of a "third" culture that overcomes the cleft between Snow's two cultures.

Snow envisioned this third culture as follows [Snow 1998, 100]:

> With good fortune, however, we can educate a large proportion of our better minds so that they are not ignorant of imaginative experience, both in the arts and in science; nor ignorant either of the endowments of applied science, of the remediable suffering of most of their fellow humans, and of the responsibilities which, once they are seen, cannot be denied.

The US-American publisher and author Brockman refers to Snow. According to Brockman [1995, 20-21], the third culture is

> founded on the realization of the import of complexity, of evolution. Very complex systems – whether organisms, brains, the biosphere, or the universe itself – were not constructed by design: all have evolved.

Accordingly, information processes originated from evolution and underwent evolution from early, rudimentary forms to the advanced forms we see today. Social science is the discontinuous continuation of natural science inasmuch as the social forms of information processes are the discontinuous continuation of natural forms of information processes.

Evolution apparently reaches ever-higher complexity which, however, is compensated for by ever-new simplicities. The points of bifurcation that are passed from stage to stage may be deciphered as symmetry breaks. Self-organisation theory therefore serves well in interpreting the "how" of the evolution of the universe (see e.g. Ebeling and Feistel [1994], Kanitscheider [2002, 467-474], Smolin [1997] as well as [2003, 201-205], Layzer [1990]).[g]

---

[g] There is evidence that the universe has been evolving for at least 13.7 billion years. Given the insights of the philosophical underpinning of self-organisation theory, it is hard to imagine a beginning like the Big Bang. For "nothing will come of nothing". As far as

Making something subject to oneself – which makes oneself a subject – is an unfolding process. It enables us to distinguish between different types of subjects according to the degree of subjectiveness they manifest. Thereby, different types of systems show different degrees of subjectiveness. The more complex a system, the more subjectiveness it displays. This holds for the ascendence from physico-chemical self-organising (dissipative) through biotic (autopoietic) to social (re-creative) systems. It is also valid for the progress of society. The course of evolution is characterised by a drift from drivenness towards end-directedness and from materiality towards formative power.

If this is so, then there is also a drift towards an increasingly sophisticated information generativity. We can apply the categorisation made above, which makes use of a new interpretation of Aristotle's four causes, to better define information. And in order to conceive of different qualities in information-generating that are due to the difference in subjectiveness self-organising systems display, we can again turn to semiotics.

According to semiotics in the sense of Charles W. Morris [1972], who recapitulated Peirce's semiotic ideas, sign relations can be viewed as having a syntactic, a semantic and a pragmatic aspect. The first refers to the relations (parts of) signs can enter with each other, the second to the relation between sign and object, and the third to the effect signs evoke. Far from defining these aspects as different from each other without integration, we can reinterpret them in the light of EST.

As noted in chapter 4, the macro-level, the systems level, is made up of three composites often ascribed to systems. These composites can be arranged in another hierarchy to reveal the fine structure of the macro-level:

---

we know, each stage in any evolutionary process is prepared by a preceding stage in that the preceding one builds the foundation for the next, irrespective of what the next stage will be like. Hence, the stage before the first stage that human science acquires knowledge of is not "nothing" but "something else" – an object of human imagination and inquiry. This is an argument for the eternity of overall evolution as put forward by prominent materialist thinkers such as Friedrich Engels ([Woods and Grant 2002, 177-222], see Fuchs [2003]).

(1) First, we know that systems are characterised by a certain organisation (determined relations of determined elements) which imparts structure. This is the bottom of the hierarchy.
(2) Second, systems are in a certain state which exhibits a property of the system. This is the next level.
(3) Third, systems show a certain behaviour *vis-à-vis* their environments, which might be made up of co-systems or other systems or might represent a suprasystem. This is the top of the hierarchy.

The structure provides possibilities for states, and a state provides possibilities for behaviour[h]. From one level to the next there is, in principle, a qualitative leap. This difference can (but need not) be bridged by a self-organisation cycle. Thus, these levels provide the prerequisites for understanding the emergence of new system structures, system states, or system behaviour.

The *signans* was defined as the self-organised order of the system. Accordingly, this order can manifest itself in how the elements are structured by the system, or in how the state of the system is, or in how the system behaves.

And we can now assign syntactics, semantics, and pragmatics accordingly: syntactics to the structure, semantics to the state, and pragmatics to the behaviour (Figure 6.11).

The sign relations are thus encapsulated:
(1) the sign-sign-relation (or part-of-sign-to-part-of-sign-relation) is the innermost one (syntactic aspect), it is embedded in
(2) the sign-sign-object-relation (semantic aspect), which is part of
(3) the sign-sign-object-subject-relation (pragmatic aspect).

The outermost relation shapes the preceding ones which, in turn, build the prerequisites of the higher levels, as is customary with nested hierarchies.

Given the evolution of self-organising systems, the appearance of signs and their syntactic, semantic and pragmatic aspects have to be differentiated into as many types as types of systems may be distinguished.

---

[h] This formulation stems from a discussion with José María Díaz Nafría.

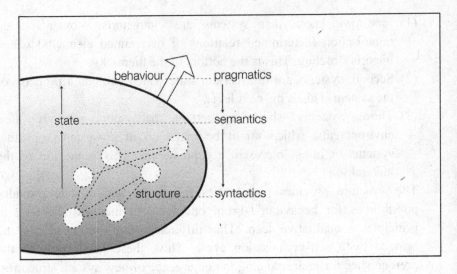

Figure 6.11. The fine structure of the macro-level of systems and of information levels as well.

At least three major quality jumps in information-generating performance become evident. These can, in turn, be subdivided, depending on the granularity we want to focus on [Haefner, 1992b]. These three types of systems have been identified here as teleomatic and self-referential, physical and chemical systems, as teleonomic and self-reproductive, biotic systems, and as teleological and self-productive, social systems. The rise of subjectiveness in the course of evolution also manifests itself in the stepwise unfolding of the structure level, state level, and behaviour level of systems. This lends an ever higher sophistication to information processes. Accordingly, an evolutionary typology of self-organising systems may show

(1) one type in which there is a short-cut between the three levels (the new structure and the new state and the new behaviour are the same, like in fluid convection),

(2) a second type in which there is still a short-cut between the upper two levels but already a differentiation into the bottom level and another one (the new structure is different from the new state, which is the same as the new behaviour, like in bird flocks) and,

(3) finally, a third type with fully-fledged three levels (like in human organisations),

This typology is consequential for information generation. The information generativity (and utility) types are pattern formation for material systems, code-making for living systems, and constitution of sense for social systems (Table 6.3).

Table 6.3. Information typology according to evolutionary system stages.

evolutionary system stage	information type
material systems	pattern formation
living systems	code-making
human systems	constitution of sense

6.2.1.1 *Pattern formation*

The most primitive sign manifestation is correlated with the most primitive kind of self-organisation processes, that is with processes by which systems (re-)structure themselves and establish a feedback loop between their micro- and macro-levels. Systems capable of this kind of self-referentiality are so-called dissipative systems. Thermodynamically speaking, they dissipate the entropy that is the byproduct of performing work when (re-)structuring. In performing work they do not only degrade energy. They also dispose of the degraded energy, which is necessary to qualify the building of the new structure as a generation of a higher order rather than a degeneration of the system. The process of (re-)structuration yields a spatial and/or temporal pattern. The pattern is the distinction that is drawn by the system.

On this stage of physical- and chemical-systems evolution there is no differentiation between the three levels of system structure, system state and system behaviour. These systems exhibit self-organisation in a dissipative thermodynamic way only: there is only one transformation function, only one cycle of self-organisation. The new structure is thus identical to the new state of the system and also to its new behaviour (see Atmanspacher and Scheingraber [1990], Atmanspacher et al. [1992], Haken [1988]). Therefore, there is no differentiation between the three semiosic aspects either. The syntactic, semantic and pragmatic aspects of information coincide. Nonetheless, the pattern *in nuce* contains all three semiosic relations.

(1) The syntactic aspect comes to the fore because the number of possible relations of elements is limited rather than unlimited. In Bénard convection cells, for example, liquid particles can move up-and-down, whereby a move upwards at a certain point in space and time excludes a move downwards at the same point.

(2) The semantic aspect comes into play because energy input into the system enables it to (re-)structure its order: the input thus becomes a signal that gives rise to the new pattern, though not completely determining it. The signal makes the state that the system adopts when forming the pattern a representation of the input. The state the system adopts "means", for the system in question, that the control parameter has exceeded a critical value beyond which it reacts with a change from heat conduction to heat convection, regardless whether the convection cells run clockwise or not.

(3) The pragmatic aspect comes up because pattern formation is tantamount to the behaviour in which the system expresses its activity *vis-à-vis* the environment. The system changes its regime of heat transport through the liquid, which is observable from the outside when the convection cells are formed.

Though analytically identifiable, the semiosic aspects in systems at this stage of evolution are not yet unfolded; they are not yet differentiated from each other, they are one. The dynamic of these systems is characterised by only one process of self-organisation. The result of this self-organisation process is only one *signans*, one sign: the pattern (Figure 6.12).

We can describe this stage of information generation as protosemiosic or proto-syntacto-semanto-pragmatic.

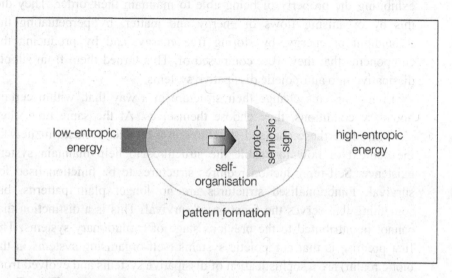

Figure 6.12. Pattern formation as information type based upon a one-levelled architecture of self-organisation.

Simply put, pattern formation is the way dissipative systems reflect (some change in the) conditions in the environment of the system. They have the ability to reflect. Simple self-organising systems in the physical and chemical domain are instances of reflective systems. Evolution, however, is a continuous process.

6.2.1.2 *Code-making*

Codes are functional. On one hand they narrow down the possibilities of the self-organised build-up of order in living systems. On the other hand they create an abundance of possibilities for survival. This is because codes regulate the decoding of signs that living systems produce when triggered by signals such that the suitability of objects/events for improving the chances of the systems' survival is reflected.

With the qualitative leap onto the stage of living-systems evolution, physical and chemical systems refined the dissipation of entropy by exhibiting the property of being able to maintain their order. They did this by organising flows of energy and matter, by perpetuating the throughput of energy, by storing free energy, and by producing the components that they were composed of. This turned them from simple dissipative into autopoietic dissipative systems.

Living systems change their structure in a way that, within certain boundary conditions, they choose themselves. At the same time, they also introduce these changed structures in a broader context, namely in the context of how to enable the structures to help maintain system existence. Self-reproduction requires structures to be functionalised for survival. Functionalised structures are no longer plain patterns, but something that serves the function of survival. This is a distinction that cannot be attributed to the previous stage of evolutionary systems. The first premise is that autopoietic systems (self-organising systems in the biotic realm) are a sophistication of dissipative systems and evolved from self-organising systems in the physical and chemical realm. This allows another self-organisation cycle to be identified, namely one that is positioned on the top of the self-organisation cycle already present and re-ontologises[i] the lower level. The postulation here is a two-levelled systems architecture. Structures that serve functions appear. This can be modeled by adding one level to the architecture of the systems (Figure 6.13). Accordingly, based upon these two self-organisation cycles – the first on a structural level, the second on a combined state/behaviour level – we can distinguish two information-generation processes.

Living systems have an aim, and they must therefore be able to assess signals from the environment and assign significance to the objects these signals come from according to the role these objects might play for maintaining their metabolism [Ayres 1988 and 1994]. Living systems show intended reactions to stimuli they perceive. This is different from saying – as the old behaviouristic dogma did – that there are strictly determined stimulus-response-relations. There is only one way to prolong the systems' "life" and improve their adaptation to the

---

[i] A word borrowed from Floridi [2007].

environment. This is to establish a meta-level from which the first level can be interpreted in the light of survival necessities and survival possibilities.

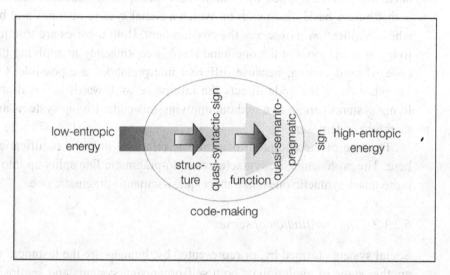

Figure 6.13. Code-making as information type based upon a two-levelled architecture of self-organisation.

This interpretation is oriented towards a goal (it is "evaluative"), requiring a rule that guides the interpretation. This rule can be denoted as a code. Usually, a code is defined as a set of rules that establish a correspondence between different sets of objects. Accordingly, DNA sequences are said to correspond to protein sequences. This is also how Marcello Barbieri [2003] uses the term when referring to the ribotype as the codemaker that mediates between the genotype and the phenotype by decoding the genotype. Note here the *caveats* that interpretation – at least basically – involves freedom of choice and that rules guide a process but never determine it or its results completely (unless the space of possibilities is restricted to one possibility only). Therefore, a code leaves enough room to move, although it channels the process into a certain direction. In the case of the genetic code, this topic is known under the label of epigenesis. The concept of code used here more closely resembles the notions advanced by cultural studies researcher Stuart Hall and sociologist Luhmann than those used in a technical sense. Hall [1997] is known for replacing the Shannon-Weaver model of communication with one that emphasises the contingency of the decoding process applied by human individuals. For Luhmann the code is that binary basic distinction on which a social system operates and by whose application it observes the environment. Both theories are true for living systems too. On the one hand there is contingency in applying the code while decoding, because different interpretations are possible. On the other hand, the code directs the interpretation towards survival and living systems cannot live without applying this code. Living systems are therefore codemakers.

The tree of the evolution of semiosic relations shows a ramification here. The protosemiosic, syntacto-semanto-pragmatic line splits up into a more quasi-syntactic one and a more quasi-semanto-pragmatic one.

6.2.1.3 *The constitution of sense*

Social systems formed by, or represented by, humans are the instances of another stage of evolution of both self-organising systems and semiosis. Note that human systems, a term which includes individuals as elements of higher-order social systems, is in sharp contrast to Luhmann's system theoretical perspective. Social systems are autopoietic systems that go beyond merely maintaining themselves: they strive for survival, but in doing so seek additional goals which they are committed to and which they have chosen on their own. They aim at realising these goals and concurrently aim at realising themselves. When they succeed they can be said to have created or re-created themselves – they are "re-creative" systems as Erich Jantsch characterised them [1987] (see also Holzkamp [1983]). They can create the conditions necessary not only for their reproduction, but also for creating themselves according to the goals they have set. They are free to restructure their environment, and with these alloplastic characteristics they are capable of restructuring themselves. As they alter their environment to suit what they themselves want to be, they exhibit even greater adaptability than nonhuman living systems. Based on these characteristics, these systems exhibit a mature distinction between the state function and the output function (for the behaviour).

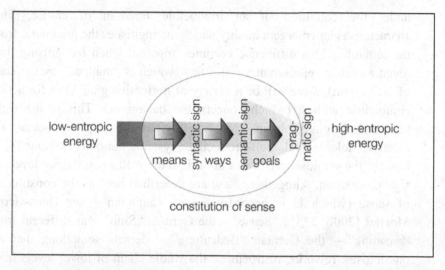

Figure 6.14. Constitution of sense as information type based upon a three-levelled architecture of self-organisation.

This means that another self-organisation cycle is added on top of the two cycles characteristic of nonhuman living systems (Figure 6.14). This new cycle reshapes the older two to yield a new unity of diverse levels: these levels may be labelled means, ways, and goals, respectively[j]. The point is to recognise that if another meta-level is added, that is, if the hierarchical structure-function distinction (characteristic of the bio-type of self-organisation) is iterated by turning the function 1 into a structure 2 for another function 2, then structure 1 has to change too because it has to serve function 2 via function 1. This represents the replacement of the structure-function relationship with the means-ways-goals relationship.

This is consequential for the information generation processes. Social systems can be classified as self-producing because, due to the self-organisation approach, they can decide which option to select when faced with a variety of behavioural options that are possible under certain circumstances. This selection takes the form of a decision and is made

---

[j] I owe this wording to systems philosopher John Collier (personal communication). In German the terms are "Mittel", "Zwecke", and "Ziele".

under the condition of an irreducible freedom of choice. That characterises an emergent quality which distinguishes the pragmatic from the semantic. This difference becomes apparent when recognising that, given a certain representation of a fact (which is "mapped" by the state of the system), there will be a variety of normative guidelines for action compatible with it (which concern the behaviour). This is the well-known Is-Ought problem. Consequently, at the stage of social, re-creative, teleological, self-productive systems, another symmetry is broken: the semiosic relations unfold into the well-known three levels of sign production. Altogether, these are described here as the constitution of sense (which is in accordance with Luhmann – see Hans-Georg Moeller [2006, 225]). "Sense" – the German "Sinn" – as different from "meaning" – the German "Bedeutung" – depicts something that re-ontologises, reworks, restructures, the whole chain of lower levels from the goals to the ways to the means in social systems. Sense is more concrete than the term "meaning", which applies to each interpretation process and hence to information processes at an evolutionary stage of all self-organising systems.

The predecessing quasi-semanto-pragmatic line branches into the two distinct but dependent pragmatic and semantic lines, which together reshape the former quasi-syntactic line as syntactic.

6.2.2 *Reflective systemic functions: the Triple-C Model*

The typology of evolutionary systems (dissipative, autopoietic, and re-creative systems) yields a typology of ever more sophisticated information-generating processes and signs (protosemiosic, quasi-semiosic, and semiosic information) during evolution. Nonetheless, it is the differentiation of different dimensions along the stepwise, hierarchical build-up of systems that lays the foundations for the functions that information processes and signs serve in any evolutionary type: the cognitive, the communicative, and the co-operative functions yield cognitive, communicative and co-operative information process or sign types [Hofkirchner and Stockinger 2003].

Subsuming metasystem transition phases and suprasystem hierarchy levels under the rubric of different "evolutionary system dimensions" helps derive an information typology according to functions.

To recapitulate those dimensions:

(1) there is the individual phase of metasystem transitions corresponding to the elementary level of suprasystem hierarchies; this is the dimension of the individual (proto-)element;

(2) there is the interactional phase of metasystem transitions corresponding to the relational level of suprasystem hierarchies; this is the dimension of interactions between individual (proto-)elements;

(3) there is the integrative phase of metasystem transitions corresponding to the systemic level of suprasystem hierarchies; this is the dimension of the integration of the individual (proto-)elements with a system.

The first dimension focuses on the internal processes of individual systems, the second on the interrelational processes of connected individual systems, and the third on processes that are external to the individual systems but internal to the meta-/suprasystem they integrate with.

Now, information generation (and utilisation) finds its functions along these dimensions and their relation to each other (Table 6.4):

(1) What transpires in the dimension of the individual (proto-)element in terms of information processes turns out to be a cognitive process. Cognition then is the individual, internal generation (and utilisation) of information by a system.

(2) What transpires in the dimension of interactions between individual (proto-)elements in terms of information processes is simply communicative processes. Communication then is the interactional, interfacial generation (and utilisation) of information by (co-)systems.

(3) And what transpires in the dimension of the integration of the individual (proto-)elements with a system in terms of information processes may be denoted as co-operative processes. Co-operation is then the collective, external generation (and utilisation) of information by (co-)systems in conjunction.

The relations of these dimensions are hierarchical. Hierarchy always means that the higher level shapes the lower one, although the higher depends on the lower. Therefore, cognition is a necessary condition for communication, and communication is a necessary condition for co-operation. Given a system of systems, co-operation of these very systems shapes their communication. This, in turn, shapes the cognition in each of them. In this way, cognition, communication and co-operation are mutually conditioned. This is the meaning of the Triple-C Model.

Table 6.4. Information function classification according to evolutionary system dimensions.

evolutionary system dimension	information function
the individual element for itself	cognition
the interaction of elements	communication
the integration of elements with the system	co-operation

Clearly, the ladder of complexity can be iterated in both directions. That which is co-operation on the higher end, turns into internal cognition of a suprasystem. And that which is cognition on the lower end, is a process and product of co-operation of subsystems.

It is important to recall that "cognition", "communication" and "co-operation" in this context are not only features of human systems but of all living systems and material systems as long as they self-organise. Cognisability, communicability and co-operability must be conceded to non-human systems too, albeit in different degrees.

In order to understand the information types according to their functionality it is helpful to examine the respective partners involved in self-organisation and to elaborate on the semiotic arguments with respect to subject-object-dialectics [Hofkirchner and Ellersdorfer, 2007].

Self-organisation of/in one system might take place

(1) either in relation to some entity that, in particular, does not qualify as a system or at least not as a currently self-organising system (a mechanical system or a self-organising system that has lost its ability to self-organise),

(2) or in relation to another self-organising system, in particular, a co-system,

(3) or in relation to the suprasystem the system takes part in.

These represent different cases of the relationship between a subject (the self-organising system) and its object, corresponding to the nature of the respective object:

(1) the object may be not specified or is a simple one,

(2) the object may be another subject of a simpler kind or of the same kind,

(3) the object may be a collective subject of which the individual subject is a part.

A simple object is an object which is no subject at all. A subject of a simpler kind than that of the subject in question features subjectiveness to a lesser degree, and a subject of the same kind features subjectiveness to the same degree. Finally, a collective subject – usually made up of a multitude of subjects of the same kind as the subject in question – is a subject of a more complex kind.

According to these cases, we can differentiate between the accompanying information generation (and utilisation) processes.

6.2.2.1 *Cognition*

In the first case, that is, in the case of engaging with whatever part of the environment from which the perturbation comes from, the most abstract form of information generation and utilisation is addressed: individual, internal, intrasubjective information. This holds particularly for engaging with a non-self-organising system, which is a simple object.

Here, the subject-object-dialectics shows opposing tendencies which might be labelled "subjection" and "objection", tending toward spiralling up in a three-step process as follows:

(1) the subject acts on the object (subjection),
(2) the object reacts (objection),
(3) the subject changes its action by taking into account the reaction of the object to its past action (new subjection).

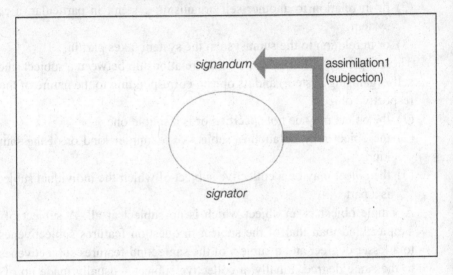

Figure 6.15. Cognition as information function based upon an engagement with an object non-specified, 1.

Regarding information or sign generation and utilisation, these steps involve assimilation – non-affordance – accommodation ("assimilation" and "accommodation" being terms introduced by Jean Piaget [1976, 1980], and "affordance" a term coined by James J. Gibson [1950, 1966, 1979] and already introduced above).

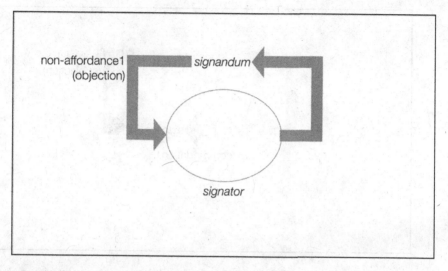

Figure 6.16. Cognition as information function based upon an engagement with an object non-specified, 2.

(1) "Assimilation" is used here to denote the informational or semiosic aspect of subjection. When subjecting the *signandum* to the system's appetence, the system attempts to render it according to how it supports meeting the system's demand. However, this is an attempt only (Figure 6.15).

(2) "Affordance" means the degree to which the *signatum* affords being subjected, "non-affordance" the degree to which it does not. When the *signatum* "objects" to – that is, resists – being being assimilated in the way attempted by the system, the whole information relation has to be overhauled (Figure 6.16).

(3) Finally, accommodation is what happens informationally or semiosically when the system attempts to adapt to what it falsely designated to serve its appetence. "Accommodation" denominates this adaptation. It works by self-organising another *signans*. Accommodation takes precedence over the next round of trying to subject the object (Figure 6.17).

Thus, information in a cognitive sense is created.

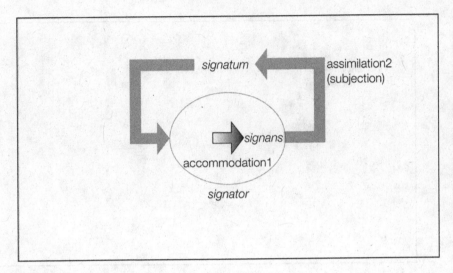

Figure 6.17. Cognition as information function based upon an engagement with an object non-specified, 3.

6.2.2.2 *Communication*

In the second case, in the case of engaging with another self-organising system, in particular with a co-system, one cognitive system is coupled to another cognitive system. Interactional, interfacial, intersubjective information generation and utilisation is the issue. It is made up of entangled intrasubjective information processes.

A number of (at least two) subjects interfere with one another. The opposing tendencies in this modified subject-object-dialectics can be called crosswise "intersubjectification" processes showing the following steps:

(1) a subject A acts on a subject B (intersubjectification by A),
(2) subject B reacts and acts on subject A (intersubjectification by B),
(3) subject A reacts and changes its action by taking into account the reaction of subject B to its past action (new intersubjectification by A).

From the informational or semiotic point of view, cognitive processes are involved in both systems and both systems are *signatores*. What they want to cognise is each other, their respective behaviours, states or structures.

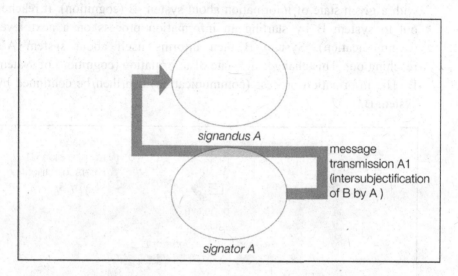

Figure 6.18. Communication as information function based upon an engagement with a self-organised system, 1.

(1) In analogy to the assimilation process, system A reaches out to system B to cognise it; it transmits a message A_1 to be received by system B (Figure 6.18).

(2) In analogy to the affordance/non-affordance reaction, system B reacts. But since it is different from a non-self-organising system, its reaction is mediated: it is irritated by message A_1 so as to produce in the course of self-organisation *signans* B_1 that, in a feedback loop, as a message is transmitted to system A (Figure 6.19).

(3) In analogy to the accommodation process, system A, when receiving message B_1, produces *signans* A_2. A_2 stands for the so signified self-organising system B and is the starting point for the transmission of a new message A_2 (which causes another irritation B_2 in system B by which system A is signified and so forth) (Figure 6.20).

System A bases its interaction with system B upon its being informed by this very system, and *vice versa*. That is, system A starts the process with a given state of information about system B (cognition). It reaches out to system B by starting an information process on a next level (communication). System B then informs itself about system A's reaching out. This changes the state of information (cognition) of system B. The information process (communication) can then be continued by system B.[k]

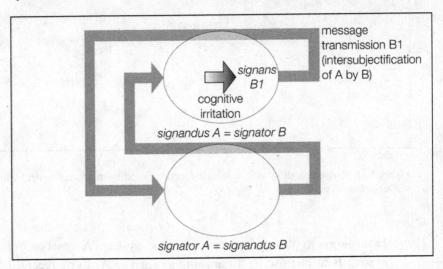

Figure 6.19. Communication as information function based upon an engagement with a self-organised system, 2.

Information in a communicative sense is created.

___

[k] My conceptualisation is on a more abstract level than Luhmann's. It therefore does not resemble his "information–message–understanding" distinction [2001]. Nonetheless, it can be added in as specifications of the cognitive irritation. I refrain from applying the term "information", though, in order not to confuse it with the generic understanding of the term used here. In the present case, both cognition and communication are instances of information. I also replace "understanding" with "irritation" to exclude interpretations that are human-centred or focussed solely on social systems.

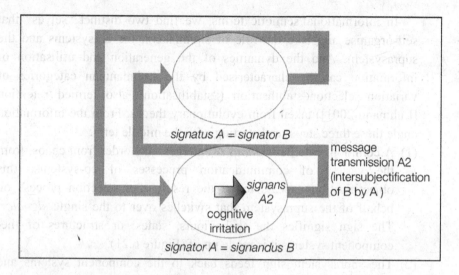

signatus A = signator B

message
transmission A2
(intersubjectification
of B by A)

signans
A2

cognitive
irritation

signator A = signandus B

Figure 6.20. Communication as information function based upon an engagement with a
self-organised system, 3.

6.2.2.3 *Co-operation*

In the third case – the engaging of co-systems with their suprasystem –
communicative systems produce integrative, external, suprasubjective
information which, in turn, informs them. A *quorum* number of co-
systems co-act and the outcome of this very co-action is a suprasystem
which, in turn, constrains and enables the co-systems' agency.

The opposing tendencies that come to the fore, seen in the context of
subject-object-dialectics, may be called "objectification" and
"subjectification". The three steps of a spiralling-up process are as
follows:
(1) fellow subjects A and B and C... act conjointly on/in a collective
    subject (objectification),
(2) the collective subject reacts (subjectification),
(3) fellow subject A or fellow subject B or fellow subject C... changes its
    contribution to the joint action by taking into account the reaction of
    the collective subject to the past action of A and B and C... (new
    objectification).

In informational/semiotic terms, we find two distinct "selves" that self-organise and take the role of signmakers: the co-systems and the suprasystem. And the dynamics of the generation and utilisation of information can be characterised by the Luhmannian categories of variation–selection–stabilisation ("stabilisation" also termed retention [Luhmann 2001]) taken from evolutionary theory. From the information angle these three steps read starting with the middle term:

(1) A common, external *signans* Δ emerges, like order from chaos, from the variety of communication processes of co-systems; this objectification appears to be the result of a selection process on behalf of the suprasystem that switches over to the single *signator*. The sign signifies the (behaviours, states, or structures of the) component systems it is made up of (Figure 6.21).

(2) The suprasystem sign feeds back to the component systems and stabilises the suprasystem inasmuch as the sign is subjectified, i.e. internalised, by the multitude of subjective co-systems (Figure 6.22).

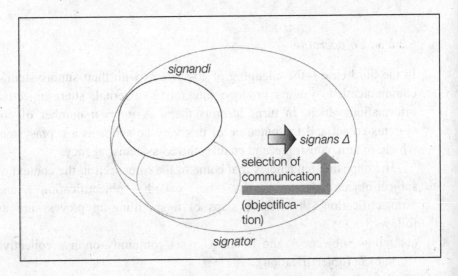

Figure 6.21. Co-operation as information function based upon an engagement with the suprasystem, 1.

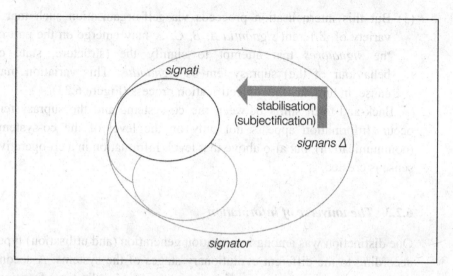

Figure 6.22. Co-operation as information function based upon an engagement with the suprasystem, 2.

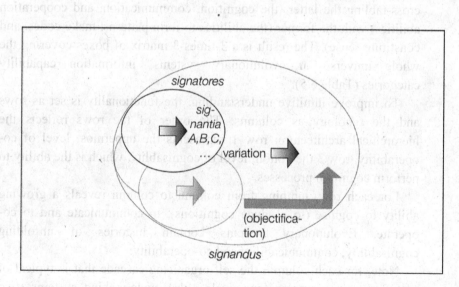

Figure 6.23. Co-operation as information function based upon an engagement with the suprasystem, 3.

(3) But this internalisation proceeds via self-organisation such that a variety of different *signantia A, B, C,* ... may emerge on the part of the *signatores* that attempt to signify the (structure, state or behaviour of the) suprasystem as *signandus*. This variation may cause, in turn, another objectification process (Figure 6.23).

Back-and-forth shifts between the co-systems and the suprasystem occur. Information appears not only on the level of the co-systems (communication), but also above that level. Information in a co-operative sense is created.

6.2.3 *The universe of information*

One distinction was among information generation (and utilisation) types according to the different evolutionary stages of the systems. A second distinction was among information generation (and utilisation) functions according to different dimensions of evolutionary systems. This enables cross-tabling the latter (the cognition, communication, and cooperation abilities) with the former (the abilities to form patterns, make codes, and constitute sense). The result is a 3-times-3 matrix of boxes covering the whole universe of evolutionary systems' information capability categories (Table 6.5).

To improve intuitive understanding, the functionality is set as rows and the typology as columns. The order of the rows reflects the hierarchical architecture: row 1 represents the uppermost level of co-operability, row 3 the bottom level of cognisablity, which is the ability to perform cognition processes.

For each row, jumping from column to column reveals a growing ability to cognise (to produce cognitions), to communicate and to co-operate. Evolutionary systems contain histories of unfolding cognisability, communicability and co-operability.

Note, for each column, the self-organising cascade that is typical of pattern-forming systems (one cycle only), code-making systems (two cycles) or sense-constituting systems (three cycles, eventually).

Table 6.5. Evolutionary systems' information capability categories. Functionality versus generativity and utility of information.

		information generativity and information utility		
		pattern-formation ability	code-making ability	sense-constituting ability
information functionality	cognisability	reflectivity	psyche	(human) consciousness
	communicability	connectivity	signalability	languageability
	co-operability	cohesiveness	organic coherency	communitarity

Each box describes the capacity of an evolutionary system to reflect, that is, to generate and utilise information structures and to take part in information processes. This capability differs based on the evolutionary type of self-organisation that characterises the system and with the systemic function performed in the system's self-organisation. Thus, with regard to cognisability, there is a lineage from primitive reflectivity in pattern formation to an advanced capability; this capability is known as psyche in code-making and ranges to (human) consciousness in sense constitution. With regard to communicability, there is a lineage from primitive connectivity in pattern formation to an advanced capability to utilise signals in code-making; this capability might be called "signalability"[l], ranging to languageability[m] in sense constitution. With regard to co-operability, there is a lineage from primitive cohesiveness in pattern formation, to an advanced capability to form organic coherency in code-making, to communitarity in sense constitution – "communitarity" denoting the specifics of human sociability. The table is

---

[l] John Collier helped me name this category.

[m] The inspiration for this term was "languaging", which I first came across in Maturana and Varela's writings.

intended to categorise all information capabilities known by science today.

This categorisation helps derive categories of information processes and structures (Table 6.6).

Table 6.6. Information categories. Functions versus types.

		evolutionary information types		
		pattern	code	sense
systemic information functions	cognition	response	flexible response	(human) reflexion
	communication	correspondences	signals	symbolic acts
	co-operation	assemblage	assignment	association

With regard to cognition, the information capability category of "reflectivity" yields the information category "response" under the rubric of pattern; "psyche" yields what can be called "flexible response" (further development of response) under the rubric of code; and (human) "consciousness" yields (human) "reflexion"[n] (further development of flexible response) under the rubric of sense. The information categories of communication are "correspondence" (mutual response), "signal" and the "symbolic act" (whereby "symbolic act" is formed in analogy to "speech act" but goes beyond conventional, that is, linguistic, human communication to also signify nonverbal, "natural" human communication such as gestures). These categories are based on

---

[n] The notion "reflexion" (with letter "x") denotes here a special variety of generic reflection, that is, a human one, which is the product of a special variety of generic reflectivity, termed "reflexivity" (a human capability of producing and using information).

connectivity, signalability and languageability, respectively. Furthermore, co-operation eventually results in "assemblage" as a pattern of cohesion; in "assignment" (further development of assemblage) as a code of organicity; and in "association" (further development of assignment) as a sense in communitarity. This is how a UTI presents a single picture of a multiplicity of information manifestations.

Summarising section 6.2, reflections in a creative universe define information as the object of praxis, reality and method of emergentist informationalism. Such reflections might be labeled "reflectivism".

Reflectivism. *Reflectivism is that strategy* vis-à-vis, *that conception of, and that research programme in information that apply the Principle of the Co-Extension of Self-Organisation and Information. The Principle of the Co-Extension of Self-Organisation and Information is based upon the Principle of the Historical and Logical Account of Information and the Saltation Principle, the Principle of Multiple Stages, and the Principle of Decentralised Context Steering.*

The **Principle of the Co-Extension of Self-Organisation and Information** states: The strategy *vis-à-vis*, the conception of, and the research programme in information tackle, view and investigate self-organisation as fulfilling the criterion of being information. Insofar as self-organisation, by mediating a self-organised order, relates a system to a perturbation, it is information realising a tripartite relationship. It realises this such that the self-organised order is the *signans* that relates the system as *signator* to the perturbation as *signandum/signatum*. Information evolves as self-organisation evolves.

In particular, the Principle of the Co-Extension of Self-Organisation and Information consists of two principles:

(I)   the **Principle of the Multi-Stage-Model Categorisation** states with regard to diachrony: Tackle, view, and investigate information along the evolutionary chain of differentiated architectures of systems as the production and utilisation of patterns, codes, and sense!

(II)   the **Principle of the Triple-C-Model Categorisation** states with regard to synchrony: Tackle, view, and investigate information across the linked intra-, inter- and suprasystemic functions as cognition, communication, and co-operation!

Reflectivism should not be confused with the outdated and degenerate variety of reflection/correspondence theory of truth in the sense of naïve realism, which is mechanistic. Reflection is a dialectic event or entity, and reflectivity a dialectic property of matter, life and human society.

The sections that follow deal with the lineages in information categories in more detail.

6.2.3.1 *Constructs for reality: from response to flexible response to reflexion*

The first step is the evolutionary lineage of cognition categories (Table 6.6, line 1). The discussion points here are the inner dynamics and architecture of evolutionary types of cognitive processes and structures against the background of what they have in common and what makes them distinct from each other.

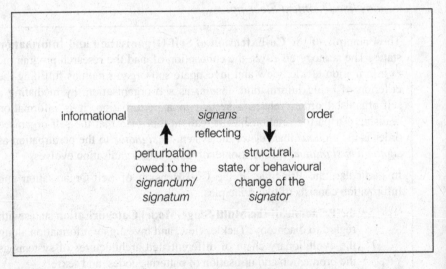

Figure 6.24. Fundamentals of cognition.

Their common features can be described by the following scheme (Figure 6.24): a process of reflection is triggered by some perturbation owed to the *signandum/signatum*, resulting in some *signans* – the order that represents some structural, state, or behavioural change of the *signator*.

Cognition is defined as follows:

Cognition. *Cognition is that informational process, or the result of that process, in which an evolutionary system relates*

(1) by carrying out a certain activity of building up order
(2) to a certain perturbation
(3) so as to realise a certain end.

Response

Pattern-forming systems reflect in a proto-semiosic way (Figure 6.25). They respond to a stimulus, and the result of that process is a response. But the response is not caused by the stimulus, as was misconceptualised in early behaviourism (and it does not gain validity in nonliving systems either). The response is mediated. It is triggered only in that the system makes a choice and produces its own kind of reflection. That sign might be called "echo", manifesting responsiveness in self-organising, nonliving matter.

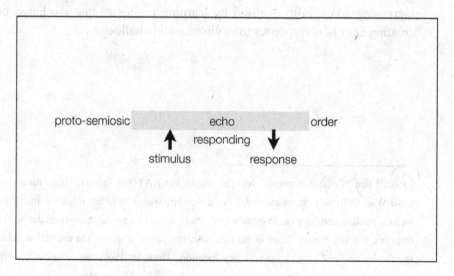

Figure 6.25. Response: echo.

Echo is the change the system shows when there is a change in the control parameter; this represents a perturbation of the system's operations. The system responds as if it would "understand" the change in the control parameter. For example, if the temperature gradient in a viscous fluid exceeds a certain (critical) value, the regime of heat transportation through the fluid switches from conduction to convection and shows characteristic hexagonal cells named after Bénard. The choice made here by the system is the clockwise or anti-clockwise rotation of the cells. Convection is more a efficient heat transportation than conduction, given the new value of the control parameter.

Thus the system exhibits cognisability in generic reflectivity. It responds to changes in its environment by self-made constructions.

Flexible response

Code-makers cognise in a more sophisticated way. By adding another cycle of self-organisation, the scope for response increases considerably and the increased flexibility makes code-making systems one order of magnitude more apt to adapt. Hence the use of the term "flexible response"[o]. The whole history of life on earth is evidence of an ever increasing adaptability realised by learning processes that are based on creating ever new responses to environmental challenges.

---

[o] Recall that "flexible response" was the title of the NATO military doctrine during the Cold War. Officially, it meant that, after a Soviet strike, retaliation might or might not include nuclear bombing or, in other words, there should be no automatism in the NATO response that the Soviet Union could calculate, and many stages of the escalation ladder should be optional at any time at any location. Thus, in 1967, the Western military abandoned the stance of behaviourism in military affairs. In the science of psychology, that stance seemed to require much more effort to be overcome by understanding the psyche as an intermediary between stimuli and responses.

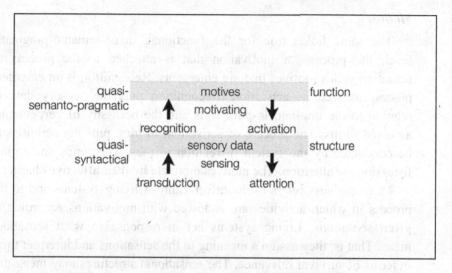

Figure 6.26. Flexible response: sensory data and motives.

By adding a second order to the first order, the cognitions produced are situated on a structural level and on a functional level and are differentiated accordingly (Figure 6.26). These codes are sensory data and motives.

Sensory data

On the structural, quasi-syntactic level, sensory data are produced by the process of sensation that includes subprocesses in opposite directions: one is the transduction of stimuli, the other is attention, the activation of receptors, the directing of sense organs towards stimuli. Note that sensing as a self-organising process is no linear cause-effect-chain but an emergent process with regard to both transduction and attention. The code of sensory data is emergent. The processes are self-organised restructurings evoked by perturbations, as Maturana and Varela termed signals from the outside, but are not strictly determined by them, and therefore not reducible to them. Restructuring on the sensory level is constrained by the space of possible relationships that elements can enter into. This space of possibilities characterises the potential of the sensorium.

Motives

The same holds true for the functional, quasi-semanto-pragmatic level: the process of motivation that is attached to the process of sensation yields motives that are emergents. Recognition is an emergent process as well as activation. Recognition puts the sensory data in relation to the implicit, in-built ends and the necessity of perpetuating assumed forms. The process of activation, in turn, puts the activities to be conducted by the system in relation to these very ends and forms, focussing the attention. The motives motivate by their affective charge.

Motivatedness refers to the effectorium of living systems and to the process in which activities are endowed with motivations according to given sensations. Living systems act in response to what sensations mean. That is, they assign a meaning to the sensations and interpret them in terms of survival relevance. The sensational structures may mean that the stimulus they represent (which is the semantic part of the sign relation) is either beneficial or detrimental to the survival of the system, or neutral (which is the pragmatic part of the sign relation). Sensations become a means to effectuation. The quasi-syntactic dimension of the new structure is supplemented by the fact that this difference makes a difference regarding the task of maintaining the system. The quasi-semanto-pragmatic dimension of codes is added.

The sensation-motivation code distinction provides the possibility to further decouple the response from the stimulus in living systems, to learn and introduce ever more intelligent systems. Yet, in contradistinction to human systems, representations and evaluations go hand in hand. An example is predator recognition. Whenever sensation is interpreted as an indication of, say, a snake or a raptor, the immediate (though context-dependent) proper activity will be motivated.

That way cognisability of code-makers is implemented as psychical constructions.

Reflexion

Sense-constituting systems have another layer added. This layer reworks the whole architecture of human cognisability and makes it another order of magnitude capable of being used for adapting the environment to the systems' self-set, explicit ends and forms. "Reflexivity" shall denote this capability, which is a feature of human thinking only (as stressed by sociologist Margaret Archer [2007]). Three varieties of reflexions can be distinguished: percepts, knowledge and wisdom as informational results of the processes of perceiving, interpreting and evaluating, respectively, on the levels of means, ways and goals, respectively (Figure 6.27).

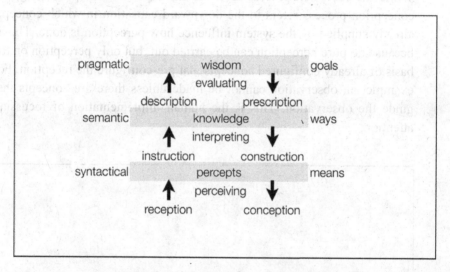

Figure 6.27. Reflexion: percepts, knowledge and wisdom.

Means, ways and goals build an upward and downward cascade in which human reflexion is generated. Human cognition is defined as a special case of cognition:

Human cognition. *Human cognition is that cognitive process, or the result of that process, in which a human system (which is labelled the "subject of recognition" or "cognitive subject"), generates or utilises information*

(1) by a certain means (which is labelled the "method of recognition" or "cognitive method")

(2) and by a certain way of signifying something (which is labelled the "object of recognition" or "cognitive object")

(3) for certain goals (which is labelled the "interest in recognition" or
*   the "cognitive interest").*

Percepts (data)

The lowest level is the syntactic level, where percepts are produced.
They are products of two opposing processes: receiving and conceiving.
While the reception process refers to the input side of perceiving, the
conception process refers to the downward causation in which concepts
already supplied by the system influence how perception is done. This is
because no pure perception can be carried out, but only perception on the
basis of already configured concepts that pre-configure the reception. For
example, an observation cannot be made unless there are concepts that
guide the observation. This is the human implementation of focussing
attention.

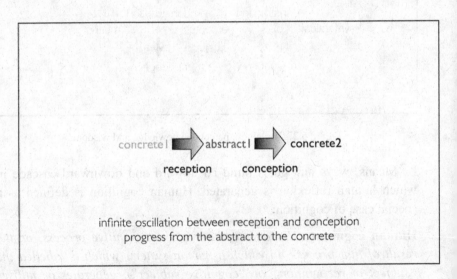

Figure 28. Condensation of percepts (data).

Percepts as unity of "recepts" and concepts[p] might be termed data, for example observational data. After disembodiment, that is, transforming them into reified physical signals outside the body, they are ready for machine data-processing procedures.

The given dialectic of perception, finally, helps condense data (Figure 28). There is an infinite oscillation between reception and conception by which progress can be made from ever new abstracts to ever new concretes.

Knowledge

The level of percepts (data) is functionalised for certain purposes, which turns it into a syntactic level. These purposes are split into two more levels, a semantic one and a pragmatic one.

The next-higher, intermediary level is the semantic level on which the creation of knowledge takes place by interpreting the percepts (data). Moments of interpretation that work in opposite directions are instruction from below (the data) and construction from above (knowledge that already exists and is applied, along with conjectures that systems come up with for the first time). Constructive moments are needed because knowledge cannot be inferred from data alone. Instructive moments are required too in order to bind knowledge back to the data. Construction influences conception, instruction is influenced by reception.

Semantics here means the relation to what knowledge is about (the *signandum/signatum*).

The dialectic of interpretation, which proceeds via the infinite oscillation between instruction and construction, progresses from relative truth to a better relative truth, thereby condensing knowledge (Figure 29).

---

[p] Some call the unity of perception and conception – which is very close to what I mean here if not the same by intention – "ception" [Talmy 1996].

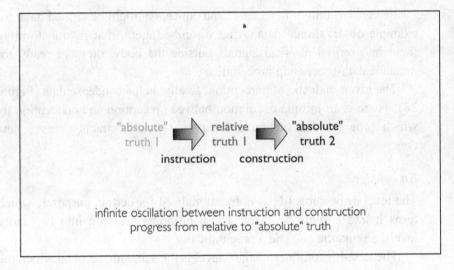

Figure 29. Condensation of knowledge.

Wisdom

The uppermost level is pragmatic. It functionalises the level below because it refers to self-set goals that determine what knowledge is (in supporting those goals). It evaluates knowledge.

The evaluation has a descriptive and a prescriptive aspect. The description starts from knowledge but is counterbalanced by prescription that orients action towards the goals (makes action pro-active) and shapes the construction of knowledge. The content of knowledge is, basically, reformulated in prescriptive terms to explicate the relation to the goals. The outcome is not only intelligence (if better actions are devised to realise the goals more elegantly), but wisdom, which is ready to question, and change, the very goals themselves. Wisdom is what enables reflexivity to enter the evolutionary stage with human cognition.

Wisdom also undergoes condensation (Figure 30). Objectivity is not the ultimate goal. By infinitely oscillating between description and prescription, the dialectic of evaluation allows for progress from objectivity towards a subjectivity that includes objectivity and incorporates it even better.

As a result of the difference between interpretation and evaluation, humans are capable of longing for wisdom, given knowledge. At the same time, things that do not make sense are possible too. This is part of the 'Is-Ought' Problem; norms cannot be concluded from facts, and the semantic cognitive level is only the prerequisite of the pragmatic cognitive level. It is, however, the ground on which the constructs build.

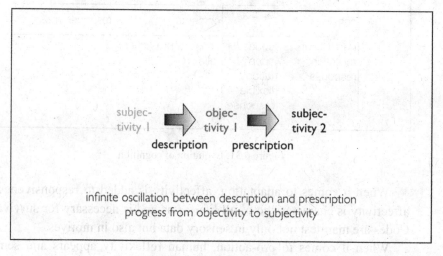

Figure 30. Condensation of wisdom.

All constructs together constitute that sense that realises consciousness.

Constructs for reality
Cognitions are evolutionary constructs.

Stagedness. Response, flexible response and reflexion, that is, patterns, codes and sense in a cognitive perspective, can be arranged according to a stage model. In this model, the evolution of cognition reveals the differentiation into ever more specific forms that build a hierarchy (Figure 6.31).

Patterns of reflection are not yet differentiated, but are evident as echo responsiveness, which is a fundamental property of self-organising matter.

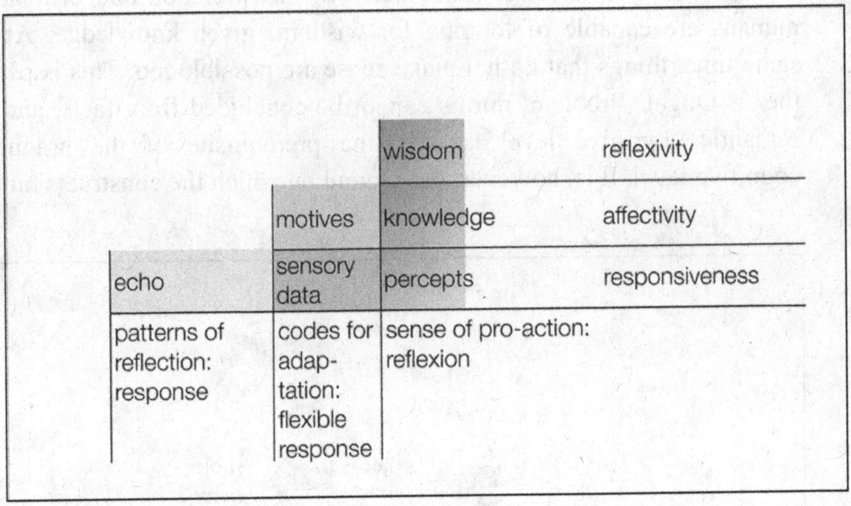

		wisdom	reflexivity
	motives	knowledge	affectivity
echo	sensory data	percepts	responsiveness
patterns of reflection: response	codes for adap- tation: flexible response	sense of pro-action: reflexion	

Figure 6.31. Evolution of cognition.

When it comes to adaptation, affectivity is added to responsiveness; affectivity is characteristic of all life forms and is necessary for survival. Codes are manifest not only in sensory data but also in motives.

When it comes to pro-action, human reflexivity appears and sense assumes the form of wisdom alongside knowledge and percepts.

Constructedness. Information emerges during all steps – during responding, during sensing and motivating, as well as during perceiving, interpreting and evaluating. This information is not predictable.

All cognitive systems exhibit subjectiveness in the sense that they construct cognitions as a product of their own activity. This activity, however, is triggered by some event or some entity in the environment. The constructs are produced to react to the challenges from outside, to respond to them without breaking down. In the case of living systems, they are produced to adapt to the challenges in order to ensure survival; in the case of human systems, to pro-actively master them to reach self-set goals. Constructs are therefore not fully arbitrary but serve a function and are, by iteration and feedback dynamics, continuously improved to better serve the function. Cognitive constructs are made for reality, for optimising the real performance of the systems.

This stance can be embraced by terming it "cognitive emergentism".

Cognitive emergentism. *Cognitive emergentism is that strategy* vis-à-vis, *that conception of, and that research programme in cognition that apply the Principle of Cognitive Emergence. The Principle of Cognitive Emergence is based upon the Principle of the Co-Extension of Self-Organisation and Information.*

The **Principle of Cognitive Emergence** states: The strategy *vis-à-vis*, the conception of, and the research programme in cognition tackle, view, and investigate cognition as a form of emergent information. Cognition emerges in qualitative leaps (I) along evolution (of cognitive system types) and (II) across hierarchies (within cognitive system types) and, at the same time, converges step by step to ever more sophisticated forms of reflection. Despite emergence, there is convergence as well.

In particular, the Principle of Cognitive Emergence consists of two principles:

(I) the **Principle of Transformational Cognitive Convergence** states with regard to diachrony: Tackle, view, and investigate cognition along a sequence that ranges from emergent patterns of reflection to emergent codes for adaptation to emergent sense of pro-action, which do not follow a deterministic line but converge, although one after the other!

(II) the **Principle of Transitional Cognitive Convergence** states with regard to synchrony: Tackle, view, and investigate cognition across the emergent level of echo, the emergent levels of sensory data and motives, and the emergent levels of percepts, knowledge and wisdom, which are not linked in a deterministic encapsulation but converge, although on top of one another.

6.2.3.2 Constructing common ground: from correspondences to signals to symbolic acts

Line 2 in Table 6.6 represents the evolutionary implementation of communicability (connectivity, signalability, and languageability). Connectivity is realised in the form of correspondence, signalability by means of signals, and languageability through symbolic acts.

The scheme underlying all three communication categories consists of intertwined cognitive processes such that reflection turns into the reflective coupling of (co-)systems, the order into coupled orders, and thus the *signans* into coupled *signantia* (Figure 6.32). The roles of communicator and communicant are introduced. Here, the communicator, who causes a perturbation that triggers a reflection by the communicant, assumes the role of the *signator* and assigns to the communicant the role of the *signandus* For the communicant the roles are reversed: the system that seeks to cognise the communicator is a *signator* and the communicator is a *signandus*.

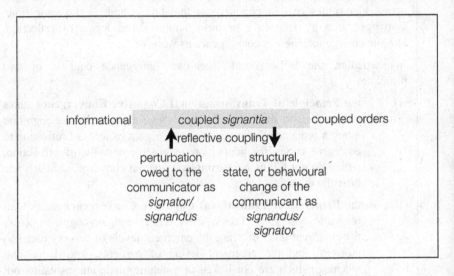

Figure 6.32. Fundamentals of communication.

Communication can then be defined:

Communication. *Communication is that informational process, or the result of that process, in which one evolutionary system relates,*

(1) by carrying out a certain activity of manifesting order
(2) referring to a certain perturbation having occurred to it previously,
(3) so as to realise a certain end,
to other evolutionary systems and vice versa.

Correspondences

In pattern formation the coupled echoes can be called "resonances" (Figure 6.33). The response of one system triggers the response of another system such that, to a greater or lesser extent, it corresponds to the first response and *vice versa*. The coupled orders of the participating systems make up the proto-semiosic resonances. Resonances demonstrate the connectedness between the systems as a result of their interaction in their alternating roles of those making signs and those being signified. They construct a basis on which they meet.

An example for resonances in the physical domain is the alignment of the rhythm of pendulums when affixed to the same wall.

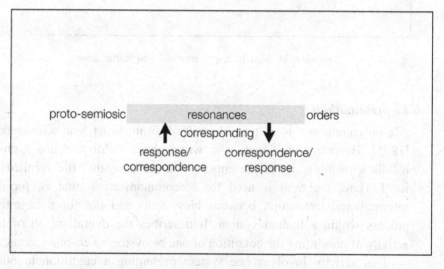

Figure 6.33. Correspondences: resonances.

Signals

On the level of living systems, structures and functions of different systems are coupled to each other. The coupled quasi-syntactic structures are termed "re-presentations", the coupled quasi-semanto-pragmatic functions are reorientations (Figure 6.34). Both are signals.

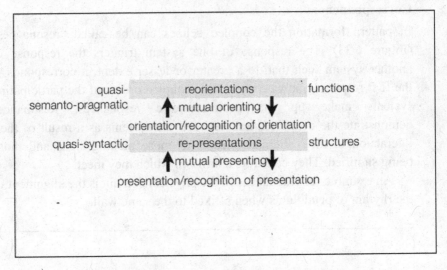

Figure 6.34. Signals: re-presentations and reorientations.

Re-presentations

"Re-presentations" is a term borrowed from Ernst von Glasersfeld [1995]. He refers to Piaget, who wrote about children: when a child recalls something, it "re-presents" or "presents again" the recalled to itself. Here, the term is used for biocommunication, that is, for the informational interaction between biosystems and not for a cognitive process within a human system. It describes the overall result of the activity of presenting the cognition of one biosystem to co-biosystems.

This activity involves one system presenting a cognition to other systems, which is countered by those very systems by recognising, to a greater or lesser extent, what is being presented. These systems reciprocate by presenting their cognitions – in particular, cognitions of the presented cognition. This, in turn, is followed by the first system recognising these presentations to a greater or lesser extent. The result is that the processes entangle and the mutual presenting yields converging re-presentations.[q]

---

[q] A model of how convergence functions that holds for semantics and pragmatics too was introduced by Fleissner and Fleissner [1998].

Reorientations

Re-presentations are topped by reorientations to which they are associated. Thus, the re-presentational signal serves the function of mutually orienting the systems. One system that is motivated to be ready for a certain activity can, by showing that motivation to other systems, influence the very motivation of these other systems. This also holds for the other systems with regard to their communication to the first system. Insofar as influences succeed, mutual orienting locks in and reorientations occur.

A well-known example is the slime mold *Dictyostelium discoideum* (see Hofkirchner and Ellersdorfer [2007, 152-154]): when in the amoeba phase, single cells begin to release the signaling molecule cAMP as soon as nutrients become scarce. This molecule can be interpreted as a presentation meaning the orientation toward leaving the single cell phase and entering the cellular phase because of nutrient deficiency. The emission of cAMP is recognised by the mediation of specific receptors of other single cells and can trigger, given certain conditions, the emission of the signaling molecule cAMP in those cells too. This is the process of mutual presenting. The cAMP molecule is the re-presentation at the structural level. Insofar as the single cells get ready to aggregate, it is a process of mutual orienting. The new state and behaviour of the single cells is the reorientation.

Another example for signalability is the process of initiating the start of a resting flock of birds. According to certain outer or inner conditions, some of the birds give, mainly by wing movements, some indication of being motivated to fly. This can be anticipated by their fellow birds, which then can react in their own way and either pass on the motive or show resistance to being infected; they may even succeed in calming down the original activists.

The communicability of biosystems is signalability – the ability to construe codes for mutual reference, which are signals.

Symbolic acts

In symbolic acts the quasi-semanto-pragmatic function is divided again into another structure and function for singling out the semantic and the pragmatic level. This makes the quasi-syntactic structure undergo a transformation to a syntactic level and fit the new process architecture of goals, ways and means. Sense in communication is produced in the interplay of the levels of the configuration, content and context of symbols (Figure 6.35).

The usage of the adjective "symbolic" hints at the outstanding feature of human communication, namely the arbitrariness and conventionality of language elements (symbols) occurring on the different levels. Symbols can be configured in a different manner (configurations), endowed with different content (contentisations) and embedded in different contexts (contextualisations).

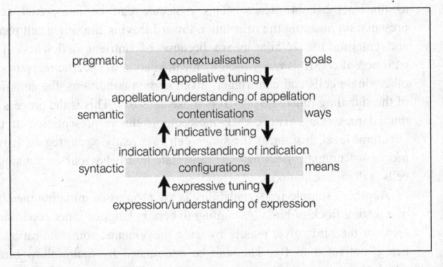

Figure 6.35. Symbolic acts: configurations, contentisations and contextualisations.

Human communication is a special case of communication. It is defined:

Human communication. *Human communication is that communicative process, or the result of that process, in which one human system (which is labelled "communicator"), tries to make intelligible,*

(1) by a certain means of demonstrating (which is labelled the "method of communication"),

(2) *a certain way of information about something (which is labelled the "object of communication")*

(3) *for certain goals (which is labelled the "interest in communication"),*

to other human systems, while the latter (which are labelled "communicants") try to grasp the configuration, content and context of symbols and engage with the first system on the same lines.

Configurations

The method of communication is processes of tuning between *ego* and *alter* in Luhmann's sense [2001] on the level of combinations of language particles on the syntactic level, by which what is presented is expressed. While *ego* expresses, *alter* attempts at understanding the expressions. In feedback loops, both tune in to each other's expressions. This comprises the whole range of possible means of presenting one's cognition, including non-verbal ones. It is expressive tuning (see Hofkirchner [2002, 183-184]).

The dialectic of expressive tuning is such that an infinite oscillation between *alter* and *ego* progresses from the expressed to the understood, consolidating the configurations (Figure 6.36).

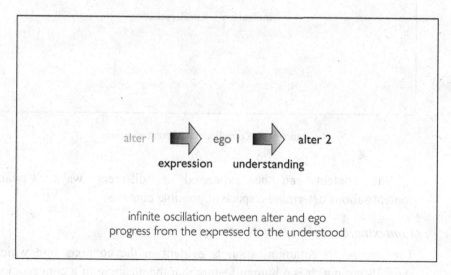

Figure 6.36. Consolidation of configurations.

Certain configurations determine a space of possible content expressions.

Contentisations

On the level of the object of communication, the content of symbols takes center stage. Again, this is a process of *alter-ego* tuning between indicating the content (what the "exchanged" cognitions are about) and understanding the content. This is indicative tuning (see Hofkirchner [2002, 185-186]). It is about semantics.

The infinite oscillation from *alter* to *ego* to *alter* yields a progress from the indicated to the understood and consolidates in a dialectic way the content of the communication (Figure 6.37).

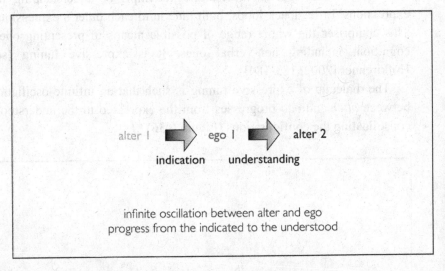

Figure 6.37. Consolidation of contentisations.

One content can be expressed in different ways. Certain contentisations determine a space of possible contexts.

Contextualisations

The interest in communication is evident in the contexts into which symbols are put. It is a human feature that the meaning of a conversation only makes sense if the relationship of *alter* and *ego* and their intentions are considered. Sometimes, it still might not make any sense.. *Alter* and *ego* appeal to each other and attempt to understand each other's appeals, the intention why *ego* wants to engage with *alter* and the relationship between *alter* and *ego*. This is appellative tuning (see Hofkirchner [2002, 186-187]).[r] It is about pragmatics.

Here as well, a dialectic of an infinite *alter-ego* oscillation consolidates the context by progressing towards the understood (Figure 6.38).

One contextualisation can embed different content.

As on all levels, a type of bargaining takes place in which *alter* and *ego* may or may not converge. Human communication leads to constructs in each human system that are influenced by the interaction with other human systems.

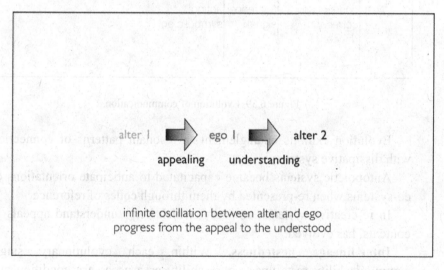

Figure 6.38. Consolidation of contextualisations.

---

[r] The appeal shows that the communicator intends to prompt the communicant to do something. This attempt at influencing and the communicator's assessment of the relation to the communicant are tightly knit together. That is why I merge both dimensions into one, in contradistinction to Schulz von Thun [1981] and others.

Constructing common ground

Communications are staged constructs that nest cognitions.

Stagedness. Correspondences in dissipative systems, signals in autopoietic systems and symbolic acts in re-creative systems are consecutive evolutionary types of communication (Figure 6.39).

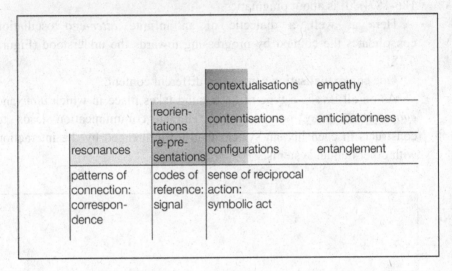

		contextualisations	empathy
	reorien-tations	contentisations	anticipatoriness
resonances	re-pre-sentations	configurations	entanglement
patterns of connection: correspon-dence	codes of reference: signal	sense of reciprocal action: symbolic act	

Figure 6.39. Evolution of communication.

Evolution exhibits entanglement in resonant patterns of connection with dissipative systems.

Autopoietic systems became capacitated to anticipate orientations of co-systems when re-presented by them through codes of reference.

In re-creative systems, empathy, which helps understand appeals in contexts, has evolved.

Inter-lineage nestedness. Within each evolutionary stage, communicability rests upon cognisability as a necessary condition and shapes the latter according to its needs, although cognisability remains relatively autonomous.

Constructedness. Resonances, re-presentations, reorientations, configurations, contentisations and contextualisations are temporary constructs because they depend on the state of interaction, intercourse or intervention. Nonetheless, communication helps construct common ground.

The above features militate in favour of communicative emergentism.

Communicative emergentism. *Communicative emergentism is that strategy* vis-à-vis, *that conception of, and that research programme in communication that apply the Principle of Communicative Emergence. The Principle of Communicative Emergence is based upon the Principle of the Co-Extension of Self-Organisation and Information.*

The **Principle of Communicative Emergence** states: The strategy *vis-à-vis*, the conception of, and the research programme in communication tackle, view, and investigate communication as a form of emergent information. Communication emerges in qualitative leaps (I) along evolution (of types of communicating systems) and (II) across hierarchies (among communicating systems of the same or similar type) and, at the same time, converges step by step to ever more sophisticated forms of coupled reflection. Despite emergence, there is convergence as well.

In particular, the Principle of Communicative Emergence consists of two principles:

(I)   the **Principle of Transformational Communicative Convergence** states with regard to diachrony: Tackle, view, and investigate communication along a sequence that ranges from emergent patterns of connection to emergent codes for reference to an emergent sense of reciprocal action, which do not follow a deterministic line but converge, although one after the other!

(II)   the **Principle of Transitional Communicative Convergence** states with regard to synchrony: Tackle, view, and investigate communication across the emergent level of resonances, the emergent levels of re-presentations and reorientations, and the emergent levels of configurations, contentisations and contextualisations, which are not linked in a deterministic encapsulation but converge, although on top of one another!

6.2.3.3 *Constructive synergy: from assemblage to assignment to association*

The last line in Table 6.6 represents the sequence of the co-operative information function along 1) the assemblage as an implementation of cohesiveness in pattern-formers, 2) the assignment as an implementation of organic coherency in code-makers, and 3) the association as an implementation of communitarity in sense-constituters.

The abstract scheme that holds for all co-operative forms of information generation and utilisation displays – instead of coupled orders such as in communication – a common order as a shared *signans* (Figure 6.40). The shared *signans* is the result of joint reflecting by a quorum of systems participating in a metasystem transition or a suprasystem hierarchy. On the one hand, the shared *signans* signifies the participating systems as *signandi/signati* viewed from the meta- or suprasystem as *signator*. On the other hand, the shared *signans* signifies the meta- or suprasystem as *signandus/signatum* from the perspective of the participating systems that play the role of *signatores*.

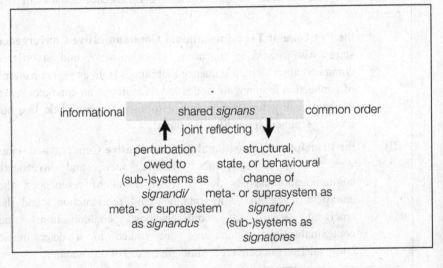

Figure 6.40. Fundamentals of co-operation.

Co-operation is defined here:

Co-operation. *Co-operation is that informational process, or the result of that process, in which a number of evolutionary systems relate*

(1) by carrying out a certain activity of building up the common order
(2) to a certain perturbation on the part of the meta- or suprasystem they participate in
(3) so as to realise a certain common end.

Assemblage

Pattern-forming systems are capable of producing cohesion. Cohesion can be defined as emergent properties that hold a system together like centripetal forces (see Collier [1986], [1988], [2003]); cohesion is a collective effort and characteristic of each collective. It involves a process of correspondences between physical, self-organising systems that, on the meta- or suprasystem level, let a *gestalt* emerge. This *gestalt*, in turn, via downward causation, profiles the structure, state or behaviour of the participating systems (Figure 6.41). *Gestalt* is the result of that very interplay of processes; this interplay is termed "assembling" here.

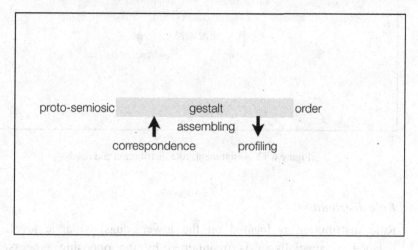

Figure 6.41. Assemblage: *gestalt*.

As an example, the particles of the fluid in the Bénard cell constellation assemble themselves to form the rotating rolls.

The assemblage is a construction to harness synergy.

Assignment

Bio-cooperation concerns the division and melding of functions between co-systems as part of an emerging metasystem or an existing suprasystem (Figure 6.42). On the one hand, there are specialisations, but on the other, special structures complement each other for a common whole.

Role distribution as structure, and needs as function, are, in the informational respect, the two codes that make up the assignment.

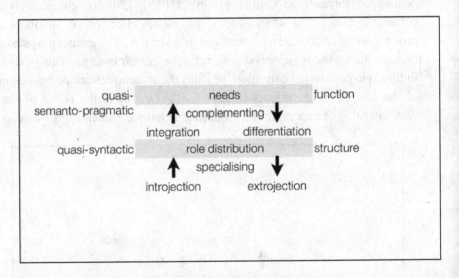

Figure 6.42. Assignment: role distribution and needs.

Role distribution

Role distribution is located on the lower, quasi-syntactic level. The balance of specialising is maintained by the opposing processes of introjection and extrojection; "introjection" means the identification with a certain role, the establishment of local rules valid for the constituent or component systems; "extrojection" refers to their performance, their contribution to the whole.

Needs

The complementing process on the higher, quasi-semanto-pragmatic level is fed by the interplay of integration and differentiation.

The ability to make assignments, that is, codes for the division of functions, is a basic feature of organisms and organic order. The division of functions is a predecessor of the division of labour in social systems. It is valid from the compartments of a single cell, through multicellular organisms, to the social organisation of insects and vertebrates. An example for such an assignment is the eventual differentiation of originally independently living amoeba-like cells of *Dictyostelium discoideum* when these gather to build a multicellular slime mould. In the slime mould, they are transformed into cells that form a stalk and die, and cells that form spores to carry on life (see Hofkirchner and Ellersdorfer [2007, 152-154]).

By way of assignment, organic coherency is able to construct more sophisticated wholes than pure physical systems can achieve.

Association

The term "meme" was introduced in analogy to the gene by Richard Dawkins [1976] to signify the unit of cultural evolution. However, it is not the ability to cause a replication *per se* that makes an individual idea a social one. And it is not the role to be mere "hosts for mind-altering strings of symbols" [Grant 1990] that is to be considered for individuals. It is correct that ideas gain a kind of supraindividual existence and that individual ideas are repercussions of socially shared ideas. But they do not evolve in accordance with a mechanistic Darwinian selection process (neither do genes).

To avoid such shortcomings of connotations of "meme", another term might be better suited: "association".[s] That term, interestingly, has two different meanings that *prima facie* signify disjoint entities or events but that can be brought together: it describes (the forming of) either a human collective or a mental connection; however, it can describe both the collective and the connection, if a human collective is comprehended as a

---

[s] My use of the term resembles certain aspects of Bruno Latour's use in his Actor-Network-Theory [2005].

group of people having something in common that is mental: a shared interest, intention, purpose. In such cases, the human collective is mentally connected. The mental connection is the informational side of material relations between humans.

In that vein, "association" denotes here social information, that is, information on the level of the meta- or suprasystem in which individual human systems participate. It evolves as a feature of social self-organisation according to a dialectic of individuals and social structures.

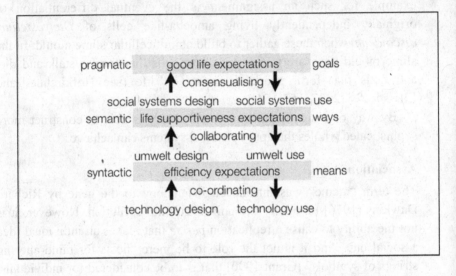

Figure 6.43. Association: efficiency, life supportiveness and good life expectations.

Sense in co-operation is constituted along an interplay of different levels (Figure 6.43). Human co-operation is merely a late evolutionary variety of co-operation and is defined as follows:

Human co-operation. *Human co-operation is that co-operative process, or the result of that process, in which a number of human systems generate or utilise social information, that is, information on the level of the social system they participate in,*

(1) by a certain means of building up a common order (which is labelled the "method of co-operation")

(2) and by certain ways of relating to a common environment (which is labelled the "object of co-operation")

(3) for certain common goals (which is labelled the "interest in co-operation").

Efficiency expectations

First, the method of co-operation is co-ordination. The division of labour requires co-ordinated activities of the human co-systems in order to increase the productive forces and their efficiency by maximising the output and minimising the input. The process of co-ordinating is realised by the interplay of designing technologies and using technologies ("technology" not solely in the sense of machines but also in the sense of any artificially devised way of carrying out an activity). The informational result of technology design and the informational medium of technology use is termed "expectations of efficiency". Within the bigger picture of human co-operation, these expectations assume a syntactic role and the role of a means. They are functionalised by higher-level expectations. "Expectation" is a term used to concretise and instantiate sense.

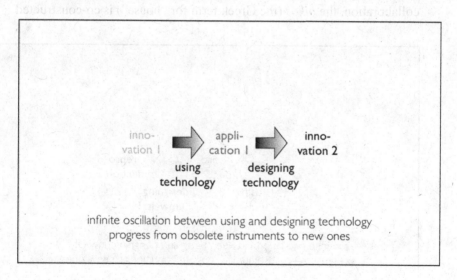

infinite oscillation between using and designing technology
progress from obsolete instruments to new ones

Figure 6.44. Co-construction of *poiesis*.

The dialectic of co-ordination is an infinite oscillation between using and designing technology, with a progress from obsolete instruments (not only reified tools but also ideational methods) to new ones. It is the co-construction of *poiesis* (the Greek "making", "producing") (Figure 6.44).

Life supportiveness expectations

Second, the object of co-operation is the environment of human co-systems, which is essentially nature, living nature. Since human systems are living systems, they strive for survival, which involves work. In work, ecosystems are addressed as life-support systems for human systems. In work, human systems collaborate. They collaboratively design and use conditions of survival, the natural *umwelt*. The process of collaborating yields what is termed here "life supportiveness expectations" as an informational product to which the semiotic label of semantics is assigned: the meaning of collaborative work is the object of co-operation.

The dialectic of collaboration shows an infinite oscillation between use and design of the *umwelt* (Figure 6.45). This is a convergent process as well because it progresses towards ever new conditions of survival. By collaboration, the *oikos* (the Greek term for "house") is co-constructed.

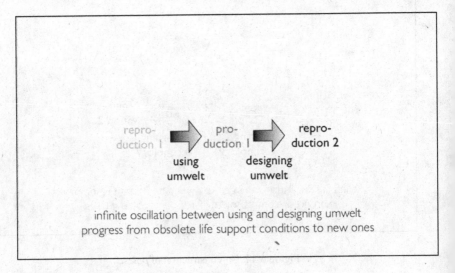

Figure 6.45. Co-construction of *oikos*.

Good life expectations

Third, the interest in co-operation is the specific human cultural dimension of living a good life, as Aristotle's *eudaimonia* denotes. Here, goals go beyond mere survival or productivity issues; the latter are only part of achieving good life. That is why goals are attached to the pragmatic level. Good life expectations are generated and utilised by consensualised actions. Consensualisation originates from designing social life (or, better: social systems) and from making use of social life (or, better: social systems). Social systems are designed and utilised because their synergy effects enable actions that human systems would otherwise be unable to carry out.

This is accompanied by a dialectic of consensualisation by which *eudaimonia* is co-constructed in an infinite oscillation between the using and the designing of social systems (Figure 6.46). Antiquated values can be replaced by new ones, which makes the process a directed one.

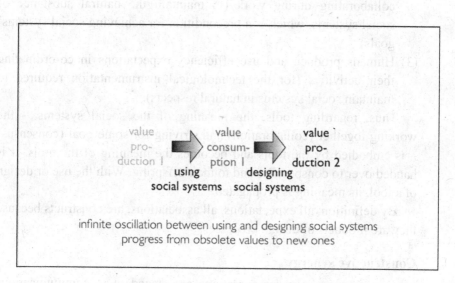

Figure 6.46. Co-construction of *eudaimonia*.

Sense, in the form of good life expectation associations, is transferred back and proliferates to the levels below. At the same time, a plethora of examples such as climate change and enduring, bloody wars exemplify that there need not be any sense at all. They demonstrate the reality of contradictory relationships between the levels and testify to the uniqueness of the species *Homo sapiens* on planet Earth. They also underline the deficiencies of not yet being successful in emancipating humans from hereditary features that are detrimental to the persistence of humanity. Nonetheless, associations are the accomplishment of communitarity.

Summarising the three-levelled architecture of associations, starting with the uppermost level, leads to the following conclusions:

(1) Humans produce and use good life expectations when consensualising their actions such that these actions make sense (for achieving social systems' goals).

(2) Humans produce and use expectations of life support while collaborating during work (to maintain the natural substance of social systems, which is a precondition for achieving social systems' goals).

(3) Humans produce and use efficiency expectations in co-ordinating their activities (for the technological instrumentation required to maintain social systems in natural respects).

Thus, regarding tools, the meaning of the social systems – the working together (collaboration) and striving for some goal (consensus) – is embodied by the tools and becomes the meaning of the tools. It is handed over to conspecifics and to their offspring. With the use or design of a tool, its meaning is propagated.

By definition, all expectations, all associations, are constructs because they are products of design processes.

Constructive synergy

Like communications, co-operations are staged, nest communications and are built by construction.

Stagedness. Collectivity is the first essential property that the evolution of self-organising systems exhibits (Figure 6.47). Patterns of cohesion show *gestalt*.

The next great innovation of nature is collective intelligence. Codes for the division of functions turn collectives into intelligent collectives that assign needs and distribute roles in a two-steps downward causation.

Eventually, with the advent of social systems, collective intelligence is topped by shared intentionality. Shared intentionality means "the participants have a joint goal in the sense that we (in mutual knowledge) do X together" [Tomasello 2009, 61]. This enables joint action. Shared intentionality causes, as a consequence of consensualisation, collaboration and co-ordination on the lower levels inherited from evolutionary history. These lower levels become the bearer of more sophisticated functions such that 1) expectations on the second level are consequences of expectations on the uppermost level for life support and 2) expectations on the lowest level are consequences of the expectations on the intermediary level for efficiency.

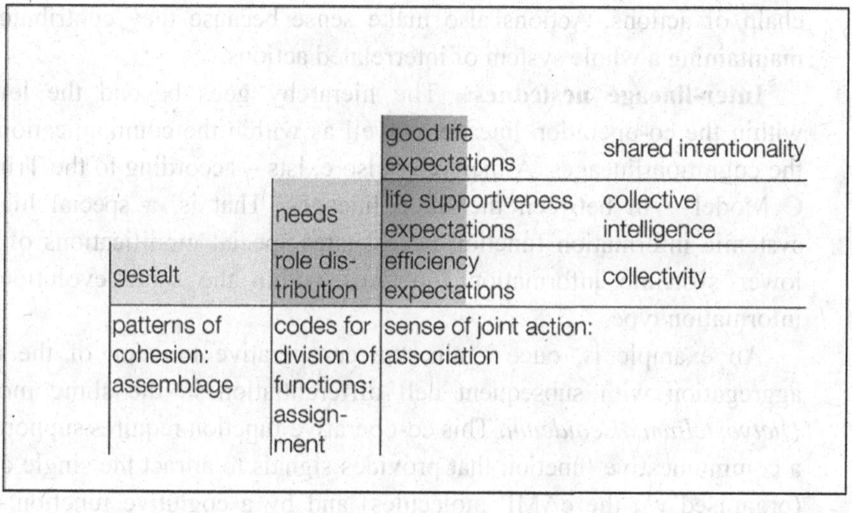

		good life expectations	shared intentionality
	needs	life supportiveness expectations	collective intelligence
gestalt	role distribution	efficiency expectations	collectivity
patterns of cohesion: assemblage	codes for division of functions: assignment	sense of joint action: association	

Figure 6.47. Evolution of co-operation.

The step to shared intentionality is a decisive leap in quality between socialising in the most advanced primates and sociability in humans. This new quality adds a new feature to evolution, which makes up a new level of semiosis. In the animal kingdom, co-operation appears in the form of certain divisions of functions. This includes, for example, the so-called "aunt behaviour" as a function in primates, or a kind of altruistic function in hoofed animals, i.e. the self-sacrificing behaviour of individual animals when the herd is chased by predators. Human co-operation takes on the form of a division of labour when at work, which serves as a model for different divided and composed actions and as a basis for different roles that individuals play in the societal context. Human co-operation reveals insight on the part of individual members into the societal context they form part of. A classic example is the hunter-beater in Aleksei N. Leontyev's activity theory [Leontyev 1981, 210-12]. Accordingly, human actions are distinct from animal behaviour in that they do not end in satisfying biotic needs but are mediated by a societal detour; humans reflect this societal detour and are aware of it. They oversee (part of) the societal context and act accordingly. Actions make sense because of their embeddedness in commonly (societally) shared designs of activity relations. This is a result of being part of a chain of actions. Actions also make sense because they contribute to maintaining a whole system of interrelated actions.

Inter-lineage nestedness. The hierarchy goes beyond the levels within the co-operation lineage as well as within the communication or the cognition lineages. A hierarchy also exists – according to the Triple-C Model – in between the three lineages. That is, a special higher systemic information function necessitates special modifications of the lower systemic information functions within the same evolutionary information type.

An example is, once again, the co-operative function of the cell aggregation with subsequent cell differentiation in the slime mould *Dictyostelium discoideum*. This co-operative function requires support by a communicative function that provides signals to attract the single cells (organised *via* the cAMP molecules) and by a cognitive function that enables recognition of the cAMP molecule.

The same holds for human co-operation, which entails human communication, which entails human cognition (see e.g. Tomasello [2009, 2008, 2000]). Expectations on the part of associations make the symbolic acts take on the form of expectations too. Human communication is fundamentally about expectations. What does *ego* expect from *alter*? What does *alter* expect *ego* to do? What does *alter* expect *ego* to expect from *alter*? Mutual expectations are formed in line with the sense that they are constituted for joint action. Human cognition, in the form of thinking, reasoning and deliberating, originated from the crisis triggered by the information overload that accompanied the increased complexity of our ancestors' social life when they acquired skills like tool-making, controlling fire, group foraging and coordinated hunting. Thinking helps deal with that chaos by creating concepts (abstract ideas that result from the generalisation of particular examples). In this regard, human cognition is distinct from animal cognition: on the basic levels, it is concept-dominated rather than sensation-focused (see [Logan 2007])[t]. Human thinking enables humans to reflect upon themselves and to reflect themselves as part of a bigger picture, that is, the social system up to society itself. Individual members of society can and do consider themselves as members of society, and they can and do consider other members as members. The social life of humans extends to societal life. The actions of members towards other members of society are mediated by this third aspect: (the structure of) society. The reflection of this is a model for all (complex) thinking. It is a model for grasping the general relationship between parts and whole, of which individual and society are merely the model instantiation.

Constructedness. The feature of being in progress all the time encompasses not only all cognitions and communications, but also all co-operative information manifestations. Synergy effects are constructed by collective processes of assembling, assigning, and constituting a sense of joint action. As they are co-constructed, they are constructed by way of synergy to improve synergy.

Such is the position of co-operative emergentism.

Co-operative emergentism. *Co-operative emergentism is that strategy vis-à-vis, that conception of, and that research programme in co-operation that apply the Principle of Co-Operative Emergence. The*

---

[t] Robert K. Logan refers here literally to "concept-based" versus "percept-based" – an opposition I do not wish to make (according to the conceptualisation presented in this book, percepts belong to the human side too).

Principle of Co-Operative Emergence is based upon the Principle of the Co-Extension of Self-Organisation and Information.

The **Principle of Co-Operative Emergence** states: The strategy *vis-à-vis*, the conception of, and the research programme in co-operation tackle, view, and investigate co-operation as a form of emergent information. Co-operation emerges in qualitative leaps (I) along evolution (of types of co-operative systems) and (II) across hierarchies (inside co-operative systems with reference to their component systems) and, at the same time, converges step by step to ever more sophisticated forms of joint reflection. Despite emergence, there is convergence as well.

In particular, the Principle of Co-Operative Emergence consists of two principles:

(I)   the **Principle of Transformational Co-Operative Convergence** states with regard to diachrony: Tackle, view, and investigate co-operation along a sequence that ranges from emergent patterns of cohesion to emergent codes for division of functions to an emergent sense of joint action, which do not follow a deterministic line but converge, although one after the other!

(II)   the **Principle of Transitional Co-Operative Convergence** states with regard to synchronicity: Tackle, view, and investigate co-operation across the emergent level of *gestalt*, the emergent levels of role distribution and needs, and the emergent levels of efficiency, life supportiveness and good life expectations, which are not linked in a deterministic encapsulation but converge, although on top of one another!

Part 3

Towards a Science for, about, and via the Information Society

Chapter 7

A Global Sustainable Information Society

To conclude, I would like to formulate some questions:

(1) Who will benefit from the networked society? Will it lower the wealth gap between nations?

(2) What will be the effect of the networked society on individual creativity?

(3) A recent poll has shown that the hope of the new millennium is for greater harmony between man and nature and amongst humans. How will the networked society affect this harmony? For me, these are not only abstract questions, but also guidelines for reflection and action.

– Ilya Prigogine: The Networked Society, 2000 –

However, these new technologies have had no such effect on the generation or acquisition of knowledge, still less on wisdom. One would hope, of course, that society would be shifting more and more towards a 'wise society', where scientifically supported data, information and knowledge would increasingly be used to make informed decisions to improve the quality of all aspects of life.

– High Level Expert Group: A European Information Society for Us All, 1997 –

According to Western thinking, since the days of Francis Bacon the role of science and technology in society is to support a better life. Today, unfortunately, the apparent effects of these inventions and innovations have come to jeopardise the originally intended aims to a degree that civilisation is at stake. This calls for overhauling Bacon's programme in the light of his ideals, and for criticizing rationality from the angle of reason [Schäfer 1993]. In this situation, reshaping science and technology is a task whose time has come.

Given that it is a shared value to improve, or at least to maintain, living conditions for the human race on this planet, the purpose of

237

scientific and technological efforts is to provide a means of coping with global problems. What is needed is self-reflexion in scientific and technological progress; the application of scientific endeavour to scientific endeavour itself, in order to redirect scientific-technological progress and to help overcome the fundamental failures of modernity; the application of research and development methods to science and technology for the purpose of their own control. Science and technology can do justice to their original purpose – to improve and simplify human life, not to mention safeguarding the conditions for survival – only when they are no longer left to pursue their seemingly natural course. Instead of being left to their own dynamics, they should be applied deliberately, after appropriate reflection and careful consideration, and should be managed with conscious control. This means executing their programme with respect to the ideals of the survival of humanity in a future that is worth living in, and constantly controlling the results of the implementation of the programme. Science must devote careful consideration to its technological consequences in society, must anticipate possible desired or undesired effects, and must carry out appropriate readjustments or reorientations.

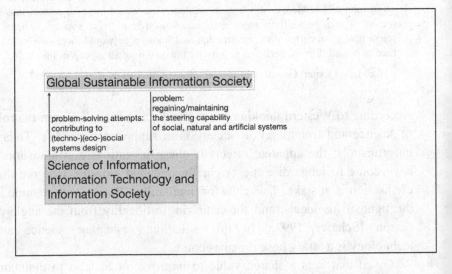

Figure 7.1. Social cybernetics as aims of a UTI-backed Science of Information.

A Science of Information will be part and parcel of the required self-reflexion of science and technology that opens the way to another, a second Modernity (Ulrich Beck, e.g. [2009], editor of the book series "Edition Zweite Moderne" with Suhrkamp, Frankfurt). Technology Assessment and social engineering are already integrated into that endeavour. A Science of Information will provide momentum to that development as soon as it has become part of the new Modernity.

A UTI may help make a Science of Information – which, furthermore, is a science of both information technology and of the information society – a science for, a science about, and a science via the information society.

By a dialectic and EST view that can be extended to a Critical Social Systems view and, beyond that, to a Critical Information Society Theory [Hofkirchner 2012], a UTI underpins the Science of Information

(1) in becoming a science for the information society in that it aims to ensure the futurability of contemporary society (see chapter section 1.3.2.1) as a social cybernetics cycle (Figure 7.1): given that complete control is out of reach, the aim is to regain the steering capability at least to the extent that a breakdown is avoided; this is to be achieved by keeping the frictions in, among and in between the social, natural and artificial subsystems of the emerging world society below the threshold of causing a breakdown. This approach provides an effective warranty that turns information society into what shall be called a "Global Sustainable Information Society (GSIS)";

(2) in its role as a science about the information society: the scope of a UTI is the ephemeral nature of information (see chapter section 1.3.2.2) by understanding the information structuration processes (reflections) as activities conducted by self-organising systems. These systems have the information structure – which is a sign – on their macro-level; on the micro-level they have the information process as the generation and utilisation of the information structure by their elements – which are informational agents themselves (Figure 7.2). These systems reside on different levels of the societal build-up – on the social level (cultural, political, economic systems),

on the environmental level (natural systems of all kinds with which social systems interact), and on the technological level (artificial systems by which social systems interact with natural systems) (Figure 7.3). The object of investigation is the nature of the reflections and their friction reduction potentialities, whose study can help optimise their synchronisation, safeguard social, environmental and technological compatibility, and integrate them with the GSIS;

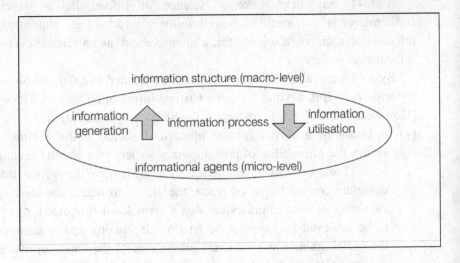

Figure 7.2. Reflections in self-organising systems as the scope of a UTI-backed Science of Information.

(3) in its instauration as a science via the information society insofar as a UTI takes advantage of complex systems thought capable of grasping the global challenges (of the information age) and insofar as it redefines the different aspects of the elephant when taking the blind men's perspective (see chapter section 1.3.2.3) as aspects that can be attained through a system of integrated system methods applied to real-world systems (Figure 7.4). Each method explains not only some quality belonging to a certain level in the holarchy of systems, but also the interrelation of the quality in question with qualities on other levels. This takes place according to a

specification scale that ranges from a dialectic systems philosophy and philosophy of information to the theory of a GSIS.

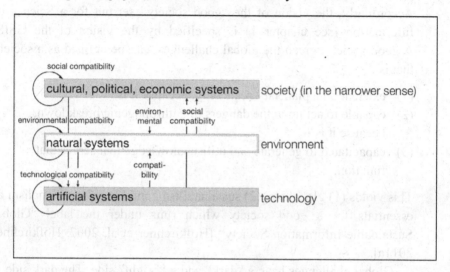

Figure 7.3. Reflections on the level of social, natural, and artificial self-organising systems as the scope of a UTI-backed Science of Information.

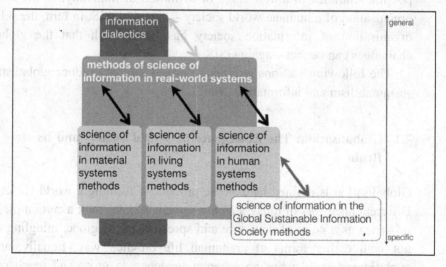

Figure 7.4. A systematics of methods as tools for a UTI-backed Science of Information.

The notion of the "GSIS" takes centre stage. It signifies the socio-political framework necessary for tackling the global challenges. Accordingly, the vision of the "good society" set out for a Science of Information (see chapter 1) is specified by the vision of the GSIS. A good society, given the global challenges, can be defined as a society that is

(1) existent on a planetary scale only, because it is
(2) capable to act upon the dangers of anthropogenic breakdown, because it is   –
(3) capacitated to generate and utilise knowledge that serves that function.

This yields (1) globalism, (2) sustainabilism, and (3) informationalism as essentials for a good society which runs under the label "Global Sustainable Information Society" [Hofkirchner et al. 2007, Hofkirchner 2011a].

Global challenges have a "dark" and a "bright" side. The dark side is the imminent danger of the breakdown of interdependent societies with the perspective of extermination of civilised human life; this is because external effects no longer remain external. The bright side marks a possible entrance to a new stage of evolution of humanity, to the self-organisation of a humane world society – namely GSIS. In turn, the self-organisation of information society has to be such that the global challenges can be met – again GSIS.

The following sections outline the processes that produce globalism, sustainabilism and informationalism.

7.1 Globalisation: The Emergence of World Society and its Brain

Globalisation is defined here as the process of forming a world society. Different ancestors of today's *Homo sapiens sapiens* took a certain place in Africa as a point of departure and spread over the globe, mingling or not with earlier forms of prehuman life on their way. Equally, any evolutionary system has an inherent tendency to grow and reach out

[Fuchs and Hofkirchner 2001, 2002a, 2002b]. What is at stake today is a metasystem transition from the component systems of humanity – a case in point are societies confined within the boundaries of nation states – to world society.

Globality characterises the state of strong interdependencies between those component systems and is, too, the result of a long evolution. It has to be complemented by a type of integration to make world society a reality. Finally, the objective factor (globality so far) has to meet a subjective factor (global consciousness) to evolve globalism.

The question posed here is: what is the contribution of the ongoing information revolution to the globalism needed for the advent of a GSIS?

From an EST point of view, society is merely another self-organising system, but one that constitutes a key step in overall evolution, namely the most sophisticated form of information generation. Computerisation and scientification may not describe the whole truth. Over and above this is another issue: will this form of social information processing, by means of electronic networking – i.e. linking humans and computers together – undergo a transformation to a new and higher level?

An underlying process may exist that bears a tendency towards ever higher human cognitive, communicative and co-operative capabilities on a planetary level. This global dimension has already been anticipated by a number of engineers, writers and academics.

There are those who argue in favour of the thesis that the spread of computer-linked telecommunications will provide the hardware of an emerging global nervous system and brain. They point out that, after the inventions of speech, writing and the printing press, the diffusion of ICTs is setting the stage for extending human collective intelligence into novel socio-technical forms. These forms might regain the high degree of inter-connectedness of bacteria [Bloom 2000]. They might also transcend the intelligence of both humans and machines today, more so than human information processing systems transcended pre-human ones [Haefner 1992a]. The introduction of each of the series of information technologies created ever closer links between individuals and groups of individuals as elements and subsystems of social systems. The same holds true for the introduction of electromagnetic communication technology and computerisation. They create interdependence at a planetary level.

Samuel Morse, who sent the first message via his electrical telegraph line from Washington to Baltimore in 1844 and succeeded in connecting Europe and North America with a durable cable in 1866, had visions of a wired world, with countries bound together by a global network of interconnected telegraph networks [Standage 1998].[a] In view of the telegraph, Nathaniel Hawthorne had one of the characters in his novel "The house of the seven gables" make the comparison of the globe with a head and brain. The paleontologist and Jesuit priest Pierre Teilhard de Chardin was not the only one to regard the "astonishing system of land, sea and air channels, the postal connections, wires, cables and radio waves, which encircle the earth more each day" as the "creation of a real nervous system of humanity, development of a common consciousness, networking of the mass of humanity," as he wrote on 6 May 1925 [1964, 61, 62; see also 1961, 117 pp. – translation from German by W.H.]. On the eve of World War II, Vladimir I. Vernadsky wrote the following [1997, 51 – translation from German by W.H.]:

> Human life has, in all its diversity, become indivisible. An event that takes place in the remotest corner of any continent or ocean has consequences, and causes reactions in a number of other places on the earth, be they great or small. The telegraph, telephone, radio, airplanes and balloons have encircled the globe. Connections are becoming ever simpler and faster. Their degree of organisation increases every year ... this process of *complete habitation of the biosphere* by humans is caused by the course of history of scientific thinking, inextricably linked with the speed of communications, the success of transport technology, the possibility of *instant* transfer of thought, and its simultaneous discussion everywhere on the planet.

Biologist and Science-and-Technology Studies expert Tom Stonier considered this process to culminate in the Internet.

> In principle, this process does not differ from the evolution of primitive nervous systems into advanced mammalian brains,

---

[a] A colleague in the audience suggested the idea of looking for Morse's ideas when I reported in a lecture, after Tom Stonier had passed away, that according to Stonier's literature research it was the novelist Hawthorne who was the first to come up with the idea of a global brain. However, Morse himself seems to have come first.

wrote Tom Stonier [1992, 105]:

> Relatively few nerve cells, relatively poorly co-ordinated, evolving into an organ consisting of trillions of cells so exquisitely co-ordinated that our understanding of how it works still eludes us. With the evolution of the global brain we are dealing with a parallel process, but at a much higher level of complexity ... Each node, rather than being a neuron, is a person comprising trillions of neurons ... coupled ... to their personal computers ... We are now dealing with the very top end of the known spectrum of intelligence.

Nontheless, it is correct to state that a change in quantity is merely a necessary precondition, but not a sufficient one, for change in quality [Fleissner and Hofkirchner 1998]. Interdependence is but a step towards integration, not integration itself. There is a qualitative leap dividing phenomena at the physiological level (that is, brain phenomena such as electrical and chemical neuronal activity) from those at the psychological level (mind phenomena such as states of consciousness and conscience). Equally, a jump is required from the interconnectivity of intelligent nodes in the global network, to the 'software' of something like a mind of global society.

Furthermore, the software to be run by the super-organism of a future world society – in order to be able to sense, interpret, and respond [Stock 1993, 80-91] – is in need of reason, more than ever before. Societal development in this phase of transition is marked by sharp discrepancies: between the practice of technically unifying the world and the social theory of world unity; between the universe of communication of nation states and the universal community of humankind (repeatedly postulated in models ever since the enlightenment); between the reality of globalisation and the ideals of humanity, evolving a global mind including self-awareness, consciousness and conscience [Richter 1992].

The question remains: will information society, beyond being capable of monitoring the manifestations of crises in the socio-economic, environmental and technological spheres, also enable humans to set civilisation on a path towards sustainable development, which is tantamount to a leap in societal self-organisation?

7.2 Sustainabilisation: Friction Mitigation – The Hidden Agenda of the Information Revolution

Sustainabilisation may be defined as the process of putting society in a position to avoid anthropogenic breakdown and safeguard a stable path of development (sustainability) by keeping social disparities below the threshold of endangering the maintenance of society.

What contribution of the current information revolution to sustainabilism – as end function of the process of sustainabilisation – is needed for the advent of a GSIS?

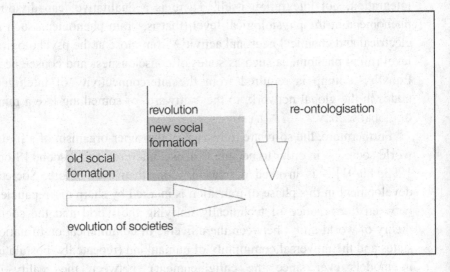

Figure 7.5. Revolution.

First, it is a revolution [Hofkirchner 2010b]. It does what revolutions in sociological terms normally do, that is, it revolutionises society. Revolutions mark changes of quality of the societal system in the course of societal evolution. Revolutions change the basis of the societal system, they form a system that differs in quality from the previous system. In doing so, the whole existing societal system is worked through and adapted accordingly to form the new system. In terms of a stage model that can be applied here for better understanding, this means that the lower stages, insofar as they build the basis of the new stage, are reworked so as to fit the emerging quality of the new whole. The new system then is permanently on the point of being formed. It might be called a "social formation". This probably covers the proper meaning of "re-ontologisation" in evolutionary systems terms (Figure 7.5). Does a revolution strengthen sustainability?

Second, it is a technological revolution. Accordingly, technology is considered to be the driving force behind revolution. "Technology" need not and, indeed, must not delimit artefacts as reified methods only, or delimit methods as ways of doing something. Rather, it needs to comprise the humans that use these ways of doing things in order to avoid blunt technological determinism. If technology triggers social change, if it is deeply intertwined with these transformations, then it would be justified to name them "techno-social formations". At least three major transformations qualify as having instigated techno-social formations: the neolithic revolution, which was a shift from nomadism to sedentariness with crop growing and cattle breeding, introduced the techno-social formation of agricultural society; the industrial revolution drew upon machine tool inventions of engineers and coupled them by transmission mechanisms with energy-providing engines like the steam engine – this yielded work machines that gave rise to the techno-social formation of industrial society; and, finally, the information revolution that is ushering in the techno-social formation of information society (Figure 7.6). Re-ontologisation occurred in each case. Each new formation subjugated the one from which it departed: the agricultural society increased the control of natural resources such as plants and animals, the industrial society has been industrialising agriculture, and the information society is informatising industry. But is informatisation boosting sustainabilisation?

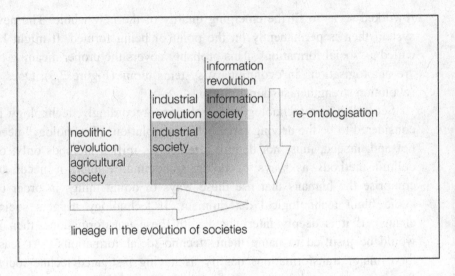

Figure 7.6. Technological revolutions.

Third, it is a scientific and technological revolution, or a scientific-technological revolution, to such an extent that, as early as in the 1950s, the British scientist and historian of science John Desmond Bernal [1954] called it that. The term immediately entered Soviet political language and ultimately became famous in 1968 and connected to the so-called Richta-report (cf. Richta [1977]). This means that technology has irreversibly become science-based. Several historical steps paved the way for this development. Though science can, in general, be seen as a response to societal needs, it started only after the neolithic revolution when, in the course of the division of labour, the creation of knowledge could take on a life on its own. Concerning the European thread of civilisation, in Antiquity, science – in the form of philosophy – was rather detached from social practice. At the turn from the 16th to the 17th century a scientific revolution occurred – a revolution in science as well as through science (cf. Volker Bialas [1978] and [1990, pp. 146]). The Copernican shift to the heliocentric planetary theory was the first act of liberation of science from being patronised by the Christian church. Francis Bacon and René Descartes followed in instaurating the new occidental science. While the industrial revolution still took place without resorting to science, industrialisation worked as a booster for it.

Technological disciplines and applied science emerged. At the turn to the 20th century, the last great independent theories were achieved by basic science. Thus, after three millennia, science was incorporated in the production process of capitalist societies. The Manhattan project that characterised the last years of World War II anticipated the so-called "Big Science". Science and technology became immediate forces of production, technology became scienticised and, vice versa, science became technologised. From that time on, the leading technology has been the computer. It delegated activities of the human brain to a machine. This, in turn, set free a variety of new technologies that would have been unable without computers. Do those technologies, however, contribute to sustainabilism?

Fourth, it is a scientific-technological revolution that provides the technology to decrease frictions appearing in the functioning of all systems. This opens up a new dimension when interpreting "the global problems [...] as frictions in the functioning of the information generation of those systems that make up world society", as the author wrote in a paper more than a decade ago (published as [Hofkirchner 2000]). Those systems are physical, biotic or social systems that are affected by the overall societal suprasystem of humans and made subsystems of it. Humans, by way of the suprasystem, are constantly engaged with those systems and they can do nothing but intervene in those systems. This intervention might be in accord with the self-organisation capacities of the systems or might be dissonant, tending to disable their self-organisation capacities. In the first case, frictions will be decreased or, at least, not increased, whereas in the second case such frictions are not decreased, eventually running the risk of damaging the system. Information technologies, knowledge-based technologies, and technologies for co-operation all can support self-organisation processes and thus ease the frictions occurring in systems they are applied to. Ultimately, social frictions, which tend to multiply and propagate throughout the subsystems of the societal suprasystem and become manifest in frictions of all kinds – social, biotic, physical – in the subsystems, can be reduced by reducing the social frictions on the suprasystem level. Sustainability can be achieved by reducing frictions.

Sustainability, denoting a society's ability to perpetuate its own development, is the most universal value to be met by a good society.

[Hofkirchner 2011a] suggests that sustainability be broken down into

(1) a social part, termed "social compatibility", which is inclusiveness and fairness – to be broken down, in turn, into equality in cultural terms, political freedom and solidarity in economic matters,

(2) an ecological part, termed "environmental compatibility", and

(3) a technological part, termed "technological compatibility", meaning a balanced relationship of new with old technologies – to be broken down, again, into usefulness, usability, efficiency, reliability, security, safety and other values.

The main argument is that society would in the long run break down and not qualify for being sustainable. This pertains not only to a society that exploits nature (e.g. reduced notions of sustainability), but also to a society that does not meet the criterion of social compatibility because of the exclusion of have-nots (who are excluded from the usage of commons regarding different societal resources); this applies particularly to exclusion of the information poor (who are excluded from imparting information). Finally, it is also valid for a society that does not abide by technology assessment that, in a participative way, includes those affected by technology.

Exclusiveness is a characteristic of societal relations of domination. Exclusion identifies societies in which some actors dominate other actors. The realisation of domination finds its predisposition in possible incongruencies in the interplay of individual and society. It is in the nature of a GSIS to be inclusive. Accordingly, the interrelation between the individual and society acknowledges their mutual enrichment. Exclusiveness denies a lasting future for society.

Today's societies lack the intelligence, logistics and organisation needed to secure their material reproduction, and to plan and carry out strategies that would set the world on a path towards sustainable development. Such development would solve problems including the use of force for political means, the gap between rich and poor (both of nations and of individuals), and damage caused by pollution and the extraction of raw materials. This obvious capacity for self-destruction is a sign that the global development of society has entered a decisive phase, a phase in which the degree of complexification and differentiation reached can be compensated for by the opposite trend of simplification and integration into a newly-created suprasystem. Contrary to evolutionary information-generating systems on the pre-human level, the type of self-organisation needed to overcome the above crises requires actions of conscious individuals. It will not emerge from technological progress alone [Laszlo 1989].

Exclusiveness can be propagated through the architecture of society and can effect or protect frictions. The underlying question remains: will the technologies of friction mitigation capacitate human agents to sustainabilise society? Will they raise awareness for mitigating frictions?

7.3 Informationalisation: ICTs for Collective Intelligence

Informationalisation is defined here as the process of raising the problem-solving capacity of world society to a level of collective intelligence that enables successfully tackling the problems arising from society's own development. Informatisation as the diffusion of information technology (ICTs) can then be defined as a means in the context of informationalisation.

What contribution of the ongoing information revolution to informationalism is needed for the advent of a GSIS?

It is true that ICTs make society increasingly responsive to information. This, however, is only a precondition for enacting collective intelligence.

Collective intelligence is a synergetic phenomenon. It designates the problem-solving capacity that results from synergetic effects of interacting intelligent agents. It was the "philosopher of the cyberspace", Pierre Lévy, who developed the concept with regard to the infosphere in 1994. His basic assumptions are these [online, without year]:

First proposition: there is a cultural evolution. Second proposition: the cultural evolution is the continuation of the biological evolution. Third proposition: the unfolding of cyberspace is the latest step of the cultural / biological evolution and the basis for future evolution. What is the role of collective intelligence in this theoretical framework? I would like to say that each step, each layer of the evolutionary continuum brings an improvement and a new realm of collective intelligence.

And he continues by saying:

Cyberspace will finally deserve its name ('piloting space' if we follow the
etymology) because it will become the driving tool (the dashboard and the wheel)
of our voyage towards a conscious biobrainsphere. The closer we get to this goal,
the wider freedom will open its space, and the more we will need to run a
multidimensional collective intelligence in real time.

Human collective intelligence is a specific form of consciousness. It
is not only the result of less friction in social terms but is, in turn, the
starting point for reducing social and other frictions, a necessary step for
a sustainable future of the suprasystem.

From this perspective, the idea of a global brain becomes functional
when viewed in the context of a possible transformation of the evolution
of consciousness into conscious evolution (Figure 7.7). Béla H. Banathy,
the advocate of social systems design, takes as his point of departure a
quotation of Jonathan Salk [1983, 112]:

... human beings now play an active and critical role not only in the process of
their own evolution but in the survival and evolution of all things.

As Banathy adds, in [2000, 203]:

If we accept this responsibility and engage creatively in the work of evolution we
shall ... be the designers of our future, we shall become the guides of our own
evolution and the evolution of life on earth and possibly beyond.

Accepting this responsibility becomes crucial because society has to
be empowered to cope with global challenges in several respects. Society
must be endowed with a means to enhance its problem-solving capacity
regarding the challenges it faces. Society must be enabled to meet the
growing demand for governance in the face of tendencies of
fragmentation, heterogenisation and disintegration.

In human systems, self-organisation is mediated via consciousness
which is the special form that information processes assume in human
systems. Conscious intervention can optimise human self-organisation as
well as self-organisation in other systems in which it intervenes, and
reduce frictions.

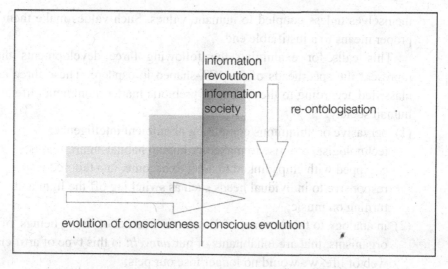

Figure 7.7. From evolution of consciousness to conscious evolution.

Thus, the information revolution might mark the beginnings of a possible transformation of the evolution of consciousness into conscious evolution, into an evolution of society that does not take place behind the backs of most members of society but is consciously and commonly carried out. ICTs inhere a potential for enhancing human collective intelligence that is required to cope with the global challenges by reducing imminent frictions. ICTs, however, can also be used to prolong exclusions and hinder the advent of a GSIS. The inclusion of stakeholders in the genesis of technology makes the design process a participatory one and ensures a discourse that will marginalise exclusions.

The vision of the GSIS as the good society is consequential for the study of ICTs and society. The vision of the GSIS does not orient itself towards a utopian "nowhere", but searches for real possibilities, i.e., possibilities anchored in reality. They are concrete and demonstrate that the search for a better society is not in vain. Those realised possibilities can be envisioned as foreshadowing a better society in the sense of philosopher Bloch.

Note that only a vision of a good society such as the GSIS provides a defensible reason for technological developments that are senseless in themselves unless coupled to humane values. Such values make them a proper means to a justifiable end.

This calls for examining the following three developments that represent the spearheads of the ICT-shaped infosphere. These three are classified according to the realms of prebiotic matter, nonhuman life and human society:

(1) pervasive or ubiquitous computing or ambient intelligence: technologists promise to make our human habitat smart, that is, equipped with chips linked to a net to become, in a tailored way, responsive to individual needs such as switching off the light and turning on music;

(2) in analogy to this Internet of Things, an internet of living beings, of organisms, that are inhabitants of our *umwelt*: in this type of artificial web of life, we would no longer lose our pets;

(3) and the internet on the level of the networked individuals of a Facebook society, a society of self-advertisement.

Each of these developments is devoid of sense. As such, each resembles the gadgets we know from our experiences as participants in the network society, unless safeguards are installed to ensure that they serve a humane purpose. Applying a GSIS perspective can, however, set the stage for the development of meaningful technologies in an evolutionary context (Figure 7.8).

The x-axis describes the dimension of virtuality and the y-axis the dimension of sociocomplexity. Virtuality means space of possibilities, sociocomplexity the complexity that arises when individual systems form a suprasystem. In the course of evolution, the space of possibilities might rise, along with the level of complexity of the systems. Physico-chemical entities, once exclusively defining the space on Planet Earth as geosphere, turn, with the rise of biota, into matter that is cycled and recycled by biota and become part of a biosphere. With the rise of human societies and the transition from biosphere to an anthropo- or sociosphere, this matter turns into inert artifacts by forming the so-called techno- or infrastructure of human societies. Finally, this matter becomes "intelligent", "smart" as "ITentities" by means of ubiquitous computing [Floridi 2007]. Biota turn, with the rise of human societies, into the living *umwelt*, and then turn, when becoming connected to the informatised infosphere, into "inforgs" [Floridi 2007]. Human societies turn, with becoming connected to the informatised infosphere, into associations networked by ICTs.

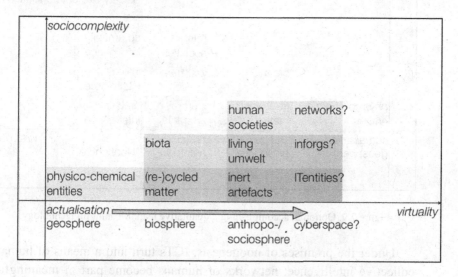

Figure 7.8. Option 1, missing a qualitative leap: Meaningless technology.

The question is whether or not the technological trends described above – and found at the respective levels – are tantamount to a real re-ontologisation of human societies. The position contended here is that only under the conditions of a GSIS is the actualisation of virtuality through ICTs tantamount to a qualitative leap onto a new level that re-ontologises the whole anthropo-/sociosphere and transforms it into a noosphere as envisaged by Teilhard de Chardin and Vernadsky (Figure 7.9).

Vernadsky considered life a geological force. The biosphere is then a result of the transformation of the geosphere(s) by life. Vernadsky's observations proved that human life on earth is a force capable of bringing about changes on the planet's surface in even shorter geological time intervals. He therefore concluded that there is a transformation of the biosphere by human work and science. This is what he called "noogenesis", the formation of the "noosphere" as another sphere [1997].

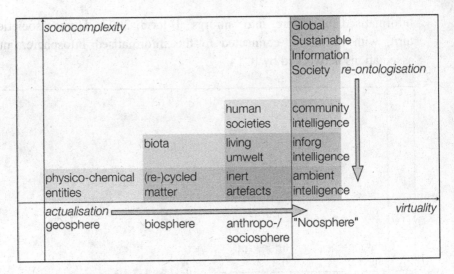

Figure 7.9. Option 2, accomplishing a qualitative leap: Meaningful technology.

Under the premises of noogenesis, ICTs turn into a means of human collective intelligence; networks of humans become part of meaningful community intelligence, inforgs become part of meaningful inforg intelligence, ITentities become part of meaningful ambient intelligence.

Without shaping ICTs to the requirements of a Global Sustainable Information Society, the technological future will be numb and dull, eventually leading to our extermination.

Bibliography

Archer, M. (2007). *Making our Way through the World, Human Reflexivity and Social Mobility* (Cambridge Universty Press, Cambridge).

Artigiani, R. (1991). Social Evolution, A Nonequilibrium Systems Model, ed. Laszlo, E., *The New Evolutionary Paradigm* (Gordon and Breach, New York), pp. 93-130.

Atmanspacher, H., Scheingraber, H. (1990). Pragmatic information and dynamical instabilities in a multimode continuous-wave dye laser, *Can. J. Phys.*, 68, pp. 728-737.

Atmanspacher, H., Kurths, J., Scheingraber, H., Wackerbauer, R., Witt, A. (1992). Complexity and Meaning in Nonlinear Dynamical Systems, *Open Systems & Inf. Dynamics*, 2/1, pp. 269-289.

Avery, T. (2000). Popper on "Social Engineering", A Classical Liberal View, *Reason Papers* 26 (Summer), St. John's University, pp. 29-38.

Ayres, R. U. (1988). Self-Organization in Biology and Economics, *Int. J. Unity of the Sc.*, 3/1, pp. 267-304.

Ayres, R. U. (1994). *Information, Entropy, and Progress* (AIP, New York).

Bacon, F. (1990). *Die Weisheit der Alten*, Herausgegeben und mit einem Essay von Philipp Rippel (Fischer, Frankfurt).

Baeyer, H. C. von (2003). *Information – The New Language of Science* (Weidenfeld and Nicolson, London).

Banathy, B. H. (2000). *Guided Evolution of Society. A Systems View* (Kluwer, New York).

Bar-Hillel, Y., and Carnap, R. (1953). Semantic Information, *Brit. J. Phil. Science* 4, pp. 147-157.

Barbieri, M. (2003). *The Organic Codes, An Introduction to Semantic Biology* (Cambridge University Press, Cambridge).

Bateson, G. (1972). *Steps to an Ecology of Mind* (Chandler, Toronto).

Baudrillard, J. (2002). "Das ist der vierte Weltkrieg", *Spiegel*, 15.1.2002, http://www.spiegel.de/kultur/gesellschaft/0,1518,177013,00.html (retrieved 05/12/2009).

Beck, U. (2009). *Macht und Gegenmacht im globalen Zeitalter* (Suhrkamp, Frankfurt).

Beniger, J. R. (1986). *The Control Revolution* (Harvard University Press, Cambridge).

Bernal, J. D. (1954). *Science in History* (Watts, London).

Bertalanffy, L. von (1950). An outline of General System Theory, *The British Journal for the Philosophy of Science*, vol. 1, no. 2, August, pp. 134-165.

Bertalanffy, L. von (1959). Human Values in a Changing World, ed Maslow A.H., *New Knowledge in Human Values* (Harper and Brothers, New York), pp. 65-74.

Bertalanffy, L. von (1965). Zur Geschichte theoretischer Modelle in der Biologie, *Studium Generale*, 18, pp. 290-298.

Bialas, V. (1978). *Wissenschaftliche und technische Revolutionen in Vergangenheit und Gegenwart* (Pahl-Rugenstein, Köln).

Bialas, V. (1990). *Allgemeine Wissenschaftsgeschichte* (Böhlau, Wien).

Bishop, R. C. (2008). Downward causation in fluid convection. *Synthese*, 160, pp. 229-248. DOI 10.1007/s11229-006-9112-2

Blind   Men   and   an   Elephant   (n.d.).   http://en.wikipedia.org/wiki/
Blind_Men_and_an_Elephant (retrieved 08/03/2008).

Blitz, D. (1992). Emergent evolution: qualitative novelty and the levels of reality (Kluwer, Dordrecht).

Bloch, E. (1967). *Das Prinzip Hoffnung*, 3 vols. (Suhrkamp, Frankfurt).

Bloom, H. (1999). Global Brain. Die Evolution sozialer Intelligenz (DVA, Stuttgart).

Brier, S. (2008). *Cybersemiotics, Why Information is not Enough* (University of Toronto Press, Toronto).

Brockman, J. (1995). *The Third Culture. Beyond the Scientific Revolution* (Simon and Schuster, New York).

Brunner, K., Klauninger, B. (2003). An Integrative Image of Causality and Emergence. eds. Arshinov, V., Fuchs, C., *Emergence, Causality, Self-Organisation* (NIA-Priroda, Moscow), pp. 23-35.

Bunge, M. (1999), Ethics and Praxiology as Technologies, *Techné*, Vol. 4, No. 4, http://scholar.lib.vt.edu/ejournals/SPT/v4n4/bunge.html#kotarbinski   (retrieved 22/11/2009).

Bunge, M. (2003). *Emergence and Convergence* (University of Toronto Press, Toronto).

Burgin, M. (2010). *Theory of Information* (World Scientific, Singapore).

Campbell, D. T. (1974). 'Downward causation' in hierarchically organized biological systems, eds. Ayala, F. J., Dobzhansky, T., *Studies in the Philosophy of Biology* (MacMillan, London), pp. 179-186.

Capurro, R. (1978). *Information* (Saur, München).

Capurro, R. (2000). What is Angeletics? http://www.capurro.de/angeletics.html (retrieved 25/02/2009).

Capurro, R. (2003). Angeletics, A Message Theory, eds. Diebner, H. H., and Ramsey, L., *Hierarchies of Communication* (ZKM, Karlsruhe), pp. 58-71.

Collen, A. (2003). *Systemic Change Through Praxis and Inquiry* (Transaction Publishers, New Brunswick, London).

Collier, J. (1986). Entropy in Evolution. *Biology and Philosophy*, 1, pp. 5-24.

Collier, J. (1988). Supervenience and Reduction in Biological Hierarchies. eds. Matthen, M., Linsky, B., *Philosophy and Biology, Canadian Journal of Philosophy Supplementary Volume* 14, pp. 209-234.

Collier, J. (2003). Hierarchical Dynamical Information Systems With a Focus on Biology. *Entropy*, 5, pp. 57-78.

Collier, J. (2008). Information in Biological Systems, eds. Adriaans, P., and Benthem, J. van, *Handbook of Philosophy of Science, Vol. 8, Philosophy of Information* (Elsevier).

Corning, P. A. (1983). The Synergism Hypothesis, A Theory of Progressive Evolution (McGraw-Hill, New York).

Corning, P. A. (1998). The Synergism Hypothesis, On the Concept of Synergy and its Role in the Evolution of Complex Systems, *J. of Social and Evolutionary Systems*, 21 (2), pp. 133-172.

Corning, P. A. (2003). *Nature's Magic – Synergy in Evolution and the Fate of Humankind* (Cambridge University Press, Cambridge).

Coulter, N. A. J., Johnson, A. L. (1982). Teleogenic Systems Theory as a Metasystem Methodology, *Proceedings of the 26th Annual Meeting of the Society for General Systems Research* (Louisville).

Csanyi, V. (1989). *Evolutionary systems and society, A general theory of life, mind and culture* (Duke University Press, Durham).

CSoI (2010). Strategic and Implementation Plan 2010–2015, http://soihub.org/resources/158/download/Strategic_Plan-final-Nov-1st.pdf (retrieved 06/03/2011).

Dawkins, R. (1976). *The selfish gene* (Oxford University Press, Oxford).

Dawkins, R. (1986). *The blind watchmaker* (Oxford University Press, Oxford).

De Vree, Johan K. (1990). Order and Disorder in the Human Universe, The Foundations of Behavioral and Social Science, 3 Vols. (Prime Presss, Bilthoven).

Doucette, D., Hofkirchner, W., Bichler, R., Raffl, C. (2007). Toward a New Science of Information, Proceedings, CODATA 2006, *Data Science Journal*, Vol. 6, 7 April, pp. S198-S205.

Dretske, F. (1981). *Knowledge and the Flow of Information* (MIT Press, Cambridge).

Ebeling, W. (1989). *Chaos, Ordnung und Information* (Urania, Berlin).

Ebeling, W., Feistel, R. (1994). *Chaos und Kosmos* (Spektrum, Heidelberg).

Eigen, M., Schuster, P. (1979). *The Hypercycle* (Springer, Berlin).

Ellersdorfer, G. (1998). Epigenetische Netzwerke, Die Emergenz "zellulärer Information" durch Selbstorganisation, eds. Fenzl, N., Hofkirchner, W., Stockinger, G., *Information und Selbstorganisation, Annäherungen an eine vereinheitlichte Theorie der Information* (Studienverlag, Innsbruck), pp. 181-201.

Érdi, P. (2008). *Complexity Explained* (Springer, Berlin).

Fenzl, N., and Hofkirchner, W. (1997). Information Processing in Evolutionary Systems. An Outline Conceptual Framework for a Unified Information Theory, ed. Schweitzer, F., *Self-Organization of Complex Structures: From Individual to Collective Dynamics, Foreword by Hermann Haken* (Gordon & Breach, London), pp. 59-70.

Feyerabend, P. (1975). *Against Method, Outline of an Anarchistic Theory of Knowledge* (NLB, London).

Fleissner, P., and Fleissner, G. (1998). Jenseits des chinesischen Zimmers, Der blinde Springer, Selbstorganisierte Semantik und Pragmatik am Computer, eds. Fenzl, N., Hofkirchner, W., Stockinger G., *Information und Selbstorganisation, Annäherungen an eine vereinheitlichte Theorie der Information* (Studienverlag, Innsbruck), pp. 325-339.

Fleissner, P., and Hofkirchner, W. (1996). Emergent Information. Towards a unified information theory. *BioSystems* 2-3(38), pp. 243–248.

Fleissner, P., and Hofkirchner, W. (1997). Actio non est reactio. An Extension of the Concept of Causality towards Phenomena of Information. *World Futures*, 3–4(49) & 1–4(50)/1997, pp. 409–427.

Floridi, L. (2003). From data to semantic information, *Entropy* 5, pp. 125-145.

Floridi, L. (2004). Outline of a Theory of Strongly Semantic Information, *Minds and Machines*, 14(2), pp. 197-222.

Floridi, L. (2005). Semantic conceptions of information, *Stanford Encyclopedia of Philosophy*, http://plato.stanford.edu/entries/information-semantic/ (retrieved 14/11/2009).

Floridi, L. (2007). A look into the future impact of ICTs on our lives, *The Information Society*, 23/1, pp. 59-64.

Floridi, L. (2009). Philosophical Conceptions of Information, ed. Sommaruga, G., *Formal Theories of Information: From Shannon to Semantic Information Theory and General Concepts of Information* (Springer, Berlin), pp. 13-53.

Foerster, H. von (1960). On Self-Organizing Systems and Their Environments, ed. Foerster, H. von, *Cybernetics of Cybernetics* (Future Systems: Minneapolis), pp. 220-230.

Foerster, H. von, Zopf, G. W. (1962), eds., *Principles of Self-Organization.* (Pergamon Press: Oxford).

Foerster, H. von (1984). Principles of Self-Organization – In a Socio-Managerial Context, eds. Ulrich, H., Probst, G. J. B., *Self-Organization and Management of Social Systems* (Springer, Berlin).

François, C. (2004), ed., *International Encyclopedia of Cybernetics and Systems*, 2 vols. (Saur, München).

Fuchs, C. (2003), Dialectical Materialism and the Self-Organisation of Matter, *Seeking Wisdom*, Vol. 1, No. 1, pp. 25-55.

Fuchs, C., and Hofkirchner, W. (2001). Theorien der Globalisierung, *Z* 48, pp. 21-34.

Fuchs, C., and Hofkirchner, W. (2002a). Globalisierung – Ein allgemeiner Prozess der Menschheitsgeschichte, *Z* 49, pp. 89-102.

Fuchs, C., and Hofkirchner, W. (2002b). Postfordistische Globalisierung, *Z* 50, pp. 152-165.

Fuchs-Kittowski, K. (1976). *Probleme des Determinismus und der Kybernetik in der molekularen Biologie* (Gustav Fischer Verlag, Jena).

Fuchs-Kittowski, K. (1997). Information – Neither Matter nor Mind, On the Essence and on the Evolutionary Stage Concept of Information, *World Futures*, Vol. 49, No.3-4, and Vol. 50, Nr. 1-4, pp. 551-570.

Gernert, D. (1996). Pragmatic information as a unifying concept, eds. Kornwachs, K., Jacoby, K., *Information, New Questions to a Multidisciplinary Concept* (Akademie Verlag, Berlin), pp. 147-162.

Gerthsen, C., Kneser, H. O., and Vogel, H. (1995). *Physik* (Springer, Berlin).

Gibbons, M., Nowotny, H. (2001). The Potential of Transdisciplinarity, eds. Thompson Klein, J., et al., *Transdisciplinarity: Joint Problem Solving among Science, Technology, and Society, An Effective Way for Managing Complexity* (Birkhäuser, Basel).

Gibson, J. J. (1950). *The perception of the visual world* (Houghton Mifflin, Boston).

Gibson, J.J. (1966). *The senses considered as perceptual systems* (Houghton Mifflin, Boston).

Gibson, J. J. (1979). *The Ecological Approach to Visual Perception* (Houghton Mifflin, Boston).

Giddens, A. (1984). *The Constitution of Society, Outline of the Theory of Structuration* (Polity Press, Cambridge).

Glasersfeld, E.v. (1995). *Radical constructivism* (Falmer, London).

Goerner, S. J. (1994). *Chaos and the Evolving Ecological Universe* (Gordon and Breach, Amsterdam).

Görnitz, T., Görnitz, B. (2002). *Der kreative Kosmos, Geist und Materie aus Information* (Spektrum, Heidelberg).

Goonatilake, S. (1991). *The evolution of information, Lineages in gene, culture and artefact* (Pinter, London).

Grant, G. (1990). Memes: Introduction. *Principia Cybernetica Web* (http://pespmc1.vub.ac.be/memin.html).

Grassberger, P. (1986). Toward a quantitative theory of selfgenerated complexity, *Intl. Journ. Theo. Phys.*, 25, 9, pp. 907-983.

Haefner, K. (1988). The Evolution of Information Processing (manuscript).

Haefner, K. (1992a). Information Processing at the Sociotechnical Level, ed. Haefner, K., *Evolution of Information Processing Systems, An Interdisciplinary Approach for a New Understanding of Nature and Society* (Springer, Berlin), pp. 307-319.

Haefner, K. (1992b) ed., Evolution of Information Processing Systems, An Interdisciplinary Approach for a New Understanding of Nature and Society (Springer, Berlin).

Haken, H. (1978). *Synergetics* (Springer, Berlin).

Haken, H. (1983). *Advanced Synergetics* (Springer, Berlin).

Haken, H. (1988). *Information and self-organization* (Springer, Berlin).

Hall, S. (1997). Encoding and Decoding, ed. Marris, P., *Media Studies* (Edinburgh University Press, Edinburgh), pp. 41-49.

Hall, A. D., Fagen, R. E. (1956). Definition of systems, *SGSR Yearbook* 1.

Halley, J. D., Winkler, D. A. (2008). Consistent concepts of Self-organization and Self-assembly, *Complexity*, 14 (2), pp. 10-17.

Hayles, N. K. (1999). *How we became postmodern* (University of Chicago Press, U.S.).

Hempel, C. G., Oppenheim, P. (1948). Studies in the Logic of Explanation, *Philosophy of Science*, 15, 2, pp. 135-175.

Heylighen, F. (1989). Causality as distinction conservation, *Cybernetics and Systems*, 20 (5) (quoted after François 2004, 84).

Heylighen, F. (1990). Autonomy and Cognition as the Maintenance and Processing of Distinctions, eds. Heylighen, F., Rosseel, E. and Demeyere, F., *Self-Steering and Cognition in Complex Systems, Toward a New Cybernetics* (Gordon and Breach, New York), pp. 89-106.

Heylighen, F. (1995). Selection of Organization at the Social Level, Obstacles and facilitators of metasystem transitions, eds. Heylighen, F., Joslyn, C., and Turchin, V. World Futures, Special Issue, The Quantum of Evolution: Toward a Theory of Metasystem Transitions. *Journal of General Evolution* Volume 45, pp. 1-4.

Heylighen, F. (1997). Towards a Gloabal Brain. Integrating Individuals into the World-Wide Electronic Network, *eds. Brandes, U., Neumann, C., Der Sinn der Sinne* (Steidl Verlag, Göttingen).

Heylighen, F. (2007). Accelerating Socio-Technological Evolution: from ephemeralization and stigmergy to the global brain, eds. Modelski, G., Devezas, T., and Thompson, W., *Globalization as an Evolutionary Process: Modeling Global Change* (Routledge, London), pp.286-335.

Hintikka, J. and P. Suppes (1970). *Information and inference* (Reidel: Dordrecht).

HLEG (1997). *A European Information Society for Us All*, http://www.ispo.cec.be/hleg/hleg.html (retrieved 08/01/2009).

Hofkirchner, W. (1995). "Information science" – an idea whose time has come, *Informatik Forum*, 3, pp. 99-106.

Hofkirchner, W. (1998). Emergence and the Logic of Explanation – An Argument for the Unity of Science, *Acta Polytechnica Scandinavica, Mathematics, Computing and Management in Engineering Series*, 91, pp. 23-30.

Hofkirchner, W. (1999), ed. *The Quest for a Unified Theory of Information*. Proceedings of the Second Conference on the Foundations of Information Science, With a Foreword by Klaus Haefner (Gordon and Breach, Amsterdam).

Hofkirchner, W. (2000). Tin hoc va xa hoi [Informatics and Society – Vietnamese], eds. Becker, J., Dang, N. D., Internet o Viet Nam va cac nuoc dang phat trien [*Internet in Viet Nam and other developing countries* – Vietnamese] (Nha Xuat Ban Khoa Hoc Va Ky Thuat, Ha Noi), pp. 73-84.

Hofkirchner, W. (2001). The hidden ontology: Real-world evolutionary systems concept as key to information science, *Emergence*, Vol. 3, No. 3, pp. 22-41.

Hofkirchner, W. (2002). *Projekt Eine Welt: Kognition – Kommunikation – Kooperation. Versuch über die Selbstorganisation der Informationsgesellschaft* (LitVerlag, Münster).

Hofkirchner, W. (2003). Homo creator in einem schöpferischen Universum. "Selbstorganisation" als Grundlage neuer Konzepte von "Natur" und "Gesellschaft", eds. Maurer, M., Höll, O., *Natur als Politikum* (RLI-Verlag, Wien), pp. 371-392.

Hofkirchner, W. (2004). Unity Through Diversity. Dialectics – Systems Thinking – Semiotics. *Trans* 15, http://www.inst.at/trans/15Nr/01_2/hofkirchner15.htm.

Hofkirchner, W. (2009). A Unified Theory of Information, An Outline, http://bitrumagora.files.wordpress.com/2010/02/uti-hofkirchner.pdf.

Hofkirchner, W. (2010a). *Twenty questions about a Unified Theory of Information* (Emergent Publications, Goodyear, Arizona).

Hofkirchner, W. (2010b). How to Design the Infosphere: the Fourth Revolution, the Management of the Life Cycle of Information, and Information Ethics as a Macroethics, *Knowledge, Technology and Policy*, Special Issue, Vol. 23, Issue 1-2, pp.177-192.

Hofkirchner, W. (2011a). ICTs for a Good Society, eds. Haftor, D., Mirijamdotter, A., *Information and Communication Technologies, Society and Human Beings, Theory and Framework, Honoring Gunilla Bradley* (Information Science Reference, Hershey, New York), pp. 434-443.

Hofkirchner, W. (2011b). Does Computing Embrace Self-Organisation?, eds. Burgin, M., Dodig-Crnkovic, G., *Information and Computation* (World Scientific, Singapore), pp. 185-202.

Hofkirchner, W. (2012). The Great Bifurcation. Information Revolution at the Crossroads (Ashgate, Farnham, Surrey, UK), forthcoming.

Hofkirchner, W., Ellersdorfer, G. (2007). Biological Information. Sign Processes in Living Systems, ed. M. Barbieri, *Biosemiotic Research Trends* (Nova Science Publishers, New York), pp. 141-155.

Hofkirchner, W., Fuchs, C., Raffl, C., Schafranek, M., Sandoval, M., Bichler, R. (2007). ICTs and Society – The Salzburg Approach. Towards a Theory for, about, and by means of the Information Society, *ICT&S Center Research Paper Series*, No. 3, Dec., http://icts.sbg.ac.at/media/pdf/pdf1490.pdf (retrieved 10/01/2008).

Hofkirchner, W., Schafranek, M. (2011). General System Theory, eds. Collier, J., Hooker, C., *Philosophy of Complexity, Chaos, and Non-Linearity, Handbook of the Philosophy of Science*, Vol. 10 (Elsevier, Amsterdam), forthcoming.

Hofkirchner, W., Stockinger, G. (2003). Towards a Unified Theory of Information. *404nOtF0und*, Vol. 1 (3), N. 24, January, http://www.facom.ufba.br/ciberpesquisa/404nOtF0und/404_24.htm (retrieved 29/05/2010).

Holland, J. W. (1975). *Adaptation in Natural and Artificial Systems* (University of Michigan Press, Ann Arbor).

Holland, J. W. (1998). *Emergence, from chaos to order* (Oxford University Press, Oxford).

Holzkamp, K. (1983). *Grundlegung der Psychologie* (Campus, Frankfurt).

Hörz, H. (1962). *Der dialektische Determinismus in Natur und Gesellschaft* (Deutscher Verlag der Wissenschaften, Berlin).

Hörz, H. (1971). *Materiestruktur* (Deutscher Verlag der Wissenschaften, Berlin).

Hörz, H. (1974). Marxistische Philosophie und Naturwissenschaften (Akademie-Verlag, Berlin).

Hörz, H. (1982). Dialectical Contradictions in Physics, eds. Marquit, E., Moran, P., Truitt, W. H., *Dialectical Contradictions: Contemporary Marxist Discussions*, (Marxist Educational Pr. Studies in Marxism Vol. 10, Minneapolis), pp. 201-222.

Hörz, H. (2009). *Materialistische Dialektik* (trafo, Berlin).

Hügin, U. (1996) Individuum, Gemeinschaft, Umwelt, Konzeption einer Theorie der Dynamik anthropogener Systeme (Lang, Bern).

Jantsch, E. (1987). Erkenntnistheoretische Aspekte der Selbstorganisation natürlicher Systeme, ed. Schmidt, S. J., *Der Diskurs des Radikalen Konstruktivismus* (Suhrkamp, Frankfurt), pp. 159-191.

Jessop, B. (1997). The Governance of Complexity and the Complexity of Governance, Preliminary Remarks on Some Problems and Limits of Economic Guidance, eds. Amin, A., Hausner, J., *Beyond Market and Hierarchy, Interactive Governance and Social Complexity* (Edward Elgar, Cheltenham).

Juarrero, A. (1999). *Dynamics in Action* (MIT, Cambridge).

Kampis, G. (1991). Self-Modifying Systems in Biology and Cognitive Science, A new framework for Dynamics, Information and Complexity (Pergamon Press, Oxford).

Kanitscheider, B. (2002). *Kosmologie* (Reclam, Stuttgart).

Kauffman, S. (1993). *The origins of order* (Oxford University Press, Oxford).

Kauffman, S., Logan, K. R., Este, R., Goebel, R., Hobill, D., and Shmulevich, I. (2008). Propagating organization: an enquiry, *Biology and Philosophy*, 23/1, pp. 27-45.

Koestler A. (1967). *The ghost in the machine* (Hutchinson, London).

Kotarbinski, T. (1965). *Praxiology* (Pergamon Press, Oxford, New York).

Krämer, S. (1988). *Symbolische Maschinen.* (Wissenschaftliche Buchgesellschaft, Darmstadt).

Kuhn, T. S. (1962). *The Structure of Scientific Revolutions* (University of Chicago Press, Chicago).

Küppers, B.-O. (1986). *Der Ursprung der biologischen Information* (Piper, München).

Küppers, B.-O. (2000). Die Strukturwissenschaften als Bindeglied zwischen Natur- und Geisteswissenschaften, ed. Küppers, B.-O., *Die Einheit der Wirklichkeit* (Wilhelm Fink, München), pp. 89-105.

Laszlo, E. (1987). *Evolution – The Grand Synthesis* (New Science Library, Boston).

Laszlo, E. (1989). *Global denken* (Horizonte, Rosenheim).

Latour, B. (2005). *Reassembling the social: an introduction to actor-network-theory* (Oxford University Press, Oxford).

Layzer, D. (1990). *Cosmogenesis, The Growth of Order in the Universe* (Oxford University Press, New York).

Lenin, W. I. (1977). *Materialismus und Empiriokritizismus* (Dietz, Berlin).

Leontyev, A. N. (1981). *Problems of the development of the mind* (Progress, Moscow).

Lévy, P. (1997). Collective Intelligence, Mankind's Emerging World in Cyberspace (Plenum Trade, New York).

Lévy, P. (-). Cyberspace as a metaevolutive step, http://www.planetwork.net/2000conf/presenters/levy_text.html (retrieved 23/04/2010).

Li, Z.-R., Tian, A.-J., and Zhang, L. (2010). *An Introduction to Theoretical Informatics* (China Science and Technology Press, Beijing).

Logan, R. (2007). *The Extended Mind, The Emergence of Language, the Human Mind and Culture* (University of Toronto Press, Toronto).

Logan, R. K. (forthcoming). *What is Information? - Propagating Organization in the Biosphere, the Symbolosphere, the Technosphere and the Econosphere*, Manuscript, http://www.physics.utoronto.ca/Members/logan/ (retrieved 3/11/2009).

Luhmann, N. (1981). *Soziologische Aufklärung*, 3 (Westdeutscher Verlag, Opladen).

Luhmann, N. (1984). *Soziale Systeme* (Suhrkamp, Frankfurt).

Luhmann, N. (2001). *Die Gesellschaft der Gesellschaft*, 2 vols. (Suhrkamp, Frankfurt).

Lukács, G. (1972). *History and class consciousness* (MIT Press, Cambridge).

Lyre, H. (1998). *Quantentheorie der Information* (Springer, Wien).

MacKay, D. M. (1969). *Information, Mechanism and Meaning* (MIT Press, Cambridge).

Mainzer, K. (1994). *Thinking in Complexity, The Complex dynamics of Matter, Mind, and Mankind* (Springer, Berlin).

Malaska, P. (1991). Economic and Social Evolution, The Transformational Dynamics Approach, ed. Laszlo, E., *The New Evolutionary Paradigm*, Gordon and Breach, New York), pp.131-156.

Maturana, H. R., Varela, F. (1980). *Autopoiesis and cognition* (Reidel, Dordrecht).

Mayr, E. (1974). Teleological and Teleonomic: A New Analysis. *Boston Studies in the Philosophy of Science*, Volume XIV (Reidel, Dordrecht), pp. 91-117.

Merton, R. K. (1936). The Unanticipated Consequences of Purposive Social Action, *American Sociological Review*, 1, pp. 894-904.

Miller, J. G. (1978). *Living Systems* (McGraw-Hill, New York).

Miller, J. G., and Miller, J. L. (1995). Applications of Living Systems Theory, *Systemic Action and Practice Research*, 8/1, pp. 19-45.

Mingers, J. (1995). Information and meaning: foundations for an intersubjective account, *Info Systems J*, 5, pp. 285-306.

Mingers, J. (1996). An Evaluation of Theories of Information with Regard to the Semantic and Pragmatic Aspects of Information Systems, *Systems Practice*, 9/3, pp. 187-209.

Mingers, J. (1997). Systems Typologies in the Light of Autopoieses: A Reconceptualization of Boulding's Hierarchy, and a Typology of Self-Referential Systems, *Syst. Res. Behav. Sci.*, Vol. 14, 303-313.

Mises, L. von (1996). *Human Action,* http://www.mises.org/humanactions.asp (retrieved 13/09/2004).

Moeller, H.-G. (2006). *Luhmann Explained, From Souls to Systems* (Open Court, Chicago).

Moreno, A. (1998). Information, causality and self-reference in natural and artificial systems, ed. Dubois, D. M., *Computing Anticipatory Systems: Proceedings Collection of American Institute of Physics,* pp. 202-206.

Moreno, A., and Ruiz-Mirazo, K. (2007). Key Issues Regarding the Origin, Nature, and Evolution of Complexity in Nature: Information as a Central Concept to Understand Biological Organization, eds. Capra, F., Juarrero, A., Sotolongo, P., and Uden, J. van, *Reframing Complexity* (ISCE, Mansfield), pp.59-77.

Morin, E. (1992). *The nature of nature* (Lang, New York).

Morin, E. (1999). *Seven Complex Lessons in Education for the Future* (UNESCO, Paris), http://unesdoc.unesco.org/images/0011/001177/117740eo.pdf		(retrieved 08/03/2008).

Morris, C. W. (1972). *Grundlagen der Zeichentheorie* (Hanser, München).

Muller, S. J. (2007). *Asymmetry: The Foundation of Information* (Springer, Berlin).

Nicolis, G., Prigogine, I., (1989). *Exploring Complexity* (Freeman, New York).

Nishigaki, T. (2007). For the Establishment of Fundamental Informatics on the Basis of Autopoiesis: Consideration on the Concept of Hierarchical Autonomous Systems, http://www.digital-narcis.org/english/FI-English-01.pdf (retrieved 12/06/2010).

Nöth, W. (2000). *Handbuch der Semiotik* (Metzler, Stuttgart).

Oyama, S. (2000). *The Ontogeny of Information. Developmental Systems and Evolution* (Duke University Press, Durham NC).

Peirce, C. S. (1983). *Phänomen und Logik der Zeichen.* (Suhrkamp, Frankfurt).

Peirce, C. S. (2000). *Semiotische Schriften,* Bd. 1, 2, 3 (Suhrkamp, Frankfurt).

Piaget, J. (1976). Die Äquilibration der kognitiven Strukturen (Klett, Stuttgart).

Piaget, J. (1980). *Abriß der genetischen Epistemologie* (Klett-Cotta, Stuttgart).

Popper, K. R. (1935). *Logik der Forschung* (Springer, Wien).

Popper, K. R. (1966). Of clouds and clocks: An approach to the problem of rationality and the freedom of man. The Arthur Holly Compton memorial lecture (Washington University, Washington).

Popper, K. R. (1972). *The poverty of historicism* (Routledge and Keagan, London).

Popper, K. R. (1990). *A world of propensities* (Thoemmes Press, Bristol).

Popper, K. R. (2005). *The open society and its enemies* (Routledge, Abingdon).

Popper, K. R., and Eccles, J. C. (1977). *The self and its brain* (Springer, Berlin).

Pouvreau, D. (2009). The dialectical tragedy of the concept of wholeness, Ludwig von Bertalanffy's biography revisited (ISCE Publishing, Goodyear).

Pouvreau, D., Drack, M. (2007). On the history of Ludwig von Bertalanffy's "General Systemology", and on its relationship to cybernetics, *International Journal of General Systems,* Vol. 36, No. 3, June, pp. 281-337.

Prigogine, I. (1980). *From being to becoming* (Freeman, San Francisco).

Richta, R. (1977). The Scientific and Technological Revolution and the Prospects of Social Development, ed. Dahrendorf, R., *Scientific-Technological Revolution, Social Aspects* (Sage, London), pp. 25-72.

Richter, E. (1992). *Der Zerfall de Welteinheit. Vernunft und Globalisierung in der Moderne* (Campus, Frankfurt).

Roederer, J. G. (2005). *Information and its Role in Nature* (Springer, Berlin).

Rosen, R. (1986). On information and complexity, eds. Casti, J. L., Karlqvist, A., *Complexity. Language and Life: Mathematical Approaches* (Springer, Berlin), pp. 174-195.

Russell, P. (1983). *The global brain, speculations on the evolutionary leap to planetary consciousness* (J. P. Turcher, Los Angeles).

Salk, J. (1983). *Anatomy of Reality: Merging of Intuition and Reason* (Columbia University Press, New York).

Salthe, S.N. (1996). *Development and Evolution* (MIT Press, Cambridge).

Sandkühler, H. J. (1990). Onto-Epistemologie, *Europäische Enzyklopädie zu Philosophie und Wissenschaften* (Meiner, Hamburg), pp. 608-615.

Sandkühler, H. J. (1991). Die Wirklichkeit des Wissens, Geschichtliche Einführung in die Epistemologie und Theorie der Erkenntnis (Suhrkamp, Frankfurt).

Schäfer, L. (1993). Das Bacon-Projekt, Von der Erkenntnis, Nutzung und Schonung der Matur (Suhrkamp, Frankfurt).

Schlemm, A. (2003). An Integated Notion of "Law", eds. Arshinov, V., Fuchs, C., *Causality, Emergence, Self-Organisation* (NIA-Priroda, Moscow), pp. 56-75.

Schluz von Thun, F. (1981). *Miteinander reden 1* (Rowohlt, Hamburg).

Seife, C. (2006). *Decoding the Universe, How the new science of information is explaining everything in the cosmos, from our brains to black holes* (Penguin, New York).

Shannon, C. E. (1948). A Mathematical Theory of Communication, *The Bell System Technical Journal*, Vol. 27, July, October, pp. 379-423, pp. 623-656.

Smolin, L. (1997). *The Life of the Cosmos* (Oxford University Press, New York).

Smolin, L. (2003). *Three Roads to Quantum Gravity, A New Understanding of Space, Time and the Universe* (Phoenix, London).

Sommaruga, G. (2009). *Formal Theories of Information. From Shannon to Semantic Information Theory and General Concepts of Information* (Springer, Berlin).

Snow, C. P. (1998). *The Two Cultures. A Second Look* (Cambridge University Press, Cambridge).

Standage, T. (1998). *The Victorian Internet. The remarkable story of the telegraph and the nineteenth century's online pioneers* (Weidenfeld and Nicolson, London).

Stock, G. (1993). *Metaman, The Merging of Humans and Machines into a Global Superorganism* (Simon and Schuster, New York).

Stokes, D. (1997). *Pasteur's Quadrant. Basic Science and Technological Innovation* (Brookings Inst., Washington D.C.).

Emergent Information

Stonier, T. (1990). *Information and the Internal Structure of the Universe, An Exploration into Information Physics* (Springer, Berlin).

Stonier, T. (1992). *Beyond Information, The Natural History of Intelligence* (Springer, Berlin).

Stonier, T. (1997). *Information and Meaning, An Evolutionary Perspective* (Springer, Berlin).

Taborsky, E. (1999). ed. *Semiosis Evolution Energy* (Shaker, Aachen).

Talmy, L. (1996). Fictive motion and change in language and perception, eds. Bloom, P., Peterson, M.A., Nadel, L. and Garrett, M., *Language and space* (MIT Press, Cambridge), pp. 211-276.

Teilhard de Chardin, P. (1961). *Die Entstehung des Menschen* (Beck, München) [French: Le groupe zoologique humain].

Teilhard de Chardin, P. (1964). *Auswahl aus dem Werk* (Walter, Freiburg, Olten) [French: La vision du passé].

Teubner, G., Willke, H. (1980). Dezentrale Kontextsteuerung im Recht intermediärer Verbände, ed. Voigt, R., *Verrechtlichung* (Athenäum, Königstein).

Thaler, R. H., Sunstein, C. R. (2008). *Nudge* (Yale University Press, New Haven).

Tomasello, M. (2000). *The cultural origins of human cognition* (Harvard University Press, Cambridge).

Tomasello, M. (2008). *Origins of Human Communication* (MIT Press, Cambridge).

Tomasello, M. (2009). *Why we cooperate* (MIT Press, Cambridge).

Toulmin, S. (1990). *Cosmopolis, The hidden agenda of modernity* (Free Press, New York).

Turchin, V., Joslyn, C. (1999). The metasystem transition, http://pespmc1.vub.ac.be/MST.html (retrieved 12/06/2010)

Ursul, A. D. (1970). *Information, Eine philosophische Studie* (Dietz Verlag, Berlin).

Van de Vijver, G., Salthe, S. N., Delpos, M. (1998). eds. □Evolutionary Systems: □Biological and Epistemological□Perspectives on Selection□and Self-organization. □ (Kluwer, Dordrecht).

Varela, F., Maturana, H., Uribe, R. (1974). Autopoiesis: The organization of living systems, its characterization and a model, *BioSystems*, Vol. 5, pp. 187-196.

Vernadskij, V.I. (1997). *Der Mensch in der Biosphäre. Zur Naturgeschichte der Vernunft* (Peter Lang, Wien).

Ward, P. (2009). *The Medea Hypothesis, Is Life on Earth Ultimately Self-Destructive?* (Princeton University Press, Princeton).

Weaver, W. (1949) The Mathematics of Communication, *Scientific American* 181.1, pp. 11-15.

Weingartner, P. (1996). Müssen wir unseren Gesetzesbegriff revidieren? ed. Weingartner, P., *Gesetz und Vorhersage* (Alber, Freiburg im Breisgau), pp. 179-222.

Weisstein, E. W. (n.d.) Attractor, *MathWorld*, A Wolfram Web Resource, http://mathworld.wolfram.com/Attractor.html (retrieved 10/12/2009).

Weizsäcker, E. U. von (1986), ed., Offene Systeme 1, Beiträge zur Zeitstruktur von Information, Entropie und Evolution (Klett-Cotta, Stuttgart).

Willke, H. (1995). *Systemtheorie III, Steuerungstheorie* (Fischer, Stuttgart).

Wilson, E. O. (1998). *Consilience: The Unity of Knowledge* (Knopf, New York).

Windelband, W. (1894). *Geschichte und Naturwissenschaft*, Rede zum Antritt des Rektorats der Kaiser-Wilhelms-Universität-Straßburg, gehalten am 1. Mai 1894, http://www.fh-augsburg.de/~harsch/germanica/Chronologie/19Jh/Windelband/ win_rede.html (retrieved 08/03/2008).

Woods, A., Grant, T. (2002). *Reason in Revolt, Dialectical Philosophy and Modern Science*, Vol. 1 (Algora Publishing, New York).

Young, A. (1974). *The Reflexive Universe* (Robert Briggs Associates, Green Leas).

Yushida, T. (2006). The Second Scientific Revolution in Capital Letters, The Informatic Turn, *triple-c*, 4/1, pp. 100-126.

Zemanek, H. (1988). Informationsverarbeitung und die Geisteswissenschaften, *Anzeiger der phil.hist. Klasse der Österr. Akademie d. Wiss.*, vol. 124, pp. 199-225.

Zimmermann, R. (2002). *Kritik der interkulturellen Vernunft* (mentis, Paderborn).

Index